MEDICINAL PLANTS OF INDIA
Volume 3

OTHER RELATED BOOKS

A Manual of Ethnobotany (2nd Ed)	*Jain, S.K.*
Betelvine Cultivation and Management of Diseases	*Chaurasia, J.P.*
Bibliography of Indian Ethnobotany, 2002	*Jain, S.K.*
Ethnic Aphrodisiac Plants	*Sood, S.K. et al.*
Ethnic Indian Plants in Cure Diabetes	*Sood, S.K. et al.*
Ethnobotany and Medicinal Plants of India & Nepal Vol. 1-2	*Singh, V. et al.*
Ethnobotany and Medicinal Plants of Indian Subcontinent	*Maheshwari, J.K.*
Ethnobotany of Rajasthan (India)	*Singh, V. et al.*
Flavouring and Fragrant Plants of Ethnic India	*Sood, S.K. et al.*
Flora of Andhra Pradesh (Vol. 1-5)	*Pullaiah, T. et al.*
Flora of Western Tribal Madhya Pradesh	*Samvatsar, S.*
Floristic Biodiversity of Barda Hills and its Surrounding	*Nagar, P.S.*
Flowering Plants of Thrissur Forest	*Sasidharan,N. et al*
Flowers and Vegetables of India	*Sud, R.K. et al.*
Glossary of Medicinal Plants used in Ayurveda	*Saroya, A.P. Singh*
Handbook of Indian Medicinal Plants	*Joshi, M.C.*
Herbs, Culinary, Medicinal, Aromatic	*Sud, R.K. et al.*
Herbs of Tribal Land, Jharkhand, India	*Jaipuriar, M.K.*
Identification of Common Indian Medicinal Plants	*Naik, V.N.*
Indian Folk Medicine & other plant-based products	*Singh, V.*
Leafy and Edible Plants of North-East India	*Kumar, S.*
Medicinal Plants of Bombay Presidency	*Agharkar, S.P.*
Medicinal Plants of North-East India	*Kumar, S.*
Monograph on Indian *Leucas* R. Br. (Dronapushpi) Lamiaceae	*Singh, V.*
Monograph on Indian Subtribe Cassiinae (Caesalpiniaceae)	*Singh, V.*
Multipurpose Plants	*Srivastava, P.K.*
Phytotherapeutic Wisdom of Indian Rural Aboriginals	*Singh, V.*
The Economic Plants of North East India	*Kumar, S.*
वनस्पति औषद्य विज्ञान	*भूत्या, रमेश कुमार*
वनस्पति शब्दकोश	*जैन, एस.के.*
कुमाऊँ हिमालय का लोकवनस्पति विज्ञान	*पाण्डे, पी.सी.*

MEDICINAL PLANTS OF INDIA

[A Guide to Ayurvedic & Ethnomedicinal uses of plants]

With

Identity, Botany, Phytochemistry,

Ayurvedic properties, Clinical & Ethnomedicinal uses

Volume - 3

DINESH JADHAV

SCIENTIFIC
PUBLISHERS

Published by

SCIENTIFIC PUBLISHERS (INDIA)
5 A, New Pali Road, P.O. Box 91
Jodhpur 342 001 (INDIA)
E-mail: info@scientificpub.com
Website : http://www.scientificpub.com

Print : 2026

ISBN: 9788172335458 (Hardbound) SET
ISBN: 9789387913073 (E-book) SET

ISBN: 9788172336011 (Hardbound) VOL-3
ISBN: 9789387913073 (E-book) VOL-3

ISBN:9788172335472 (Hardbound) VOL-2
ISBN: 9789387913073 (E-book) VOL-2

ISBN: 9788172335465 (Hardbound) VOL-1
ISBN: 9789387913073 (E-book) VOL-1

Printed in India

Dedicated to

The Memory of My Grandmother

Late Mrs. ***Kanchan Devi Jadhav***

PREFACE

Medicinal plants belong to the earliest known health care products that have been used by the mankind. In India, the earliest reference to the medicinal value of plants appears in *Rigveda*, in which a brief reference to the healing property of plants has been made. However, it is in the *Ayurveda* that definite properties and uses of drugs and drug yielding plants have been discussed. In fact, Ayurvedic medicine had its origin from the works of *Ayurveda*. The later works of *Charak* and *Susruta*, namely *Charak-Samhita* and *Susruta Samhita* added invaluable knowledge to the science of medicinal plants.

Most of the drug plants are found in the tropics growing in wild condition and are mainly used by herb doctors and Ayurvedic vaidyas, who refer to them as *jari-butis*. In the modern age with the fast emergence of western modern medicine and synthetic drugs, the importance of plants and herbs started declining as more and more attention was paid to the development of synthetic drugs. However, with these synthetic drug came the problems of side effects and other clinical complications; the reason in the increasing awareness about the limitations of the synthetic drugs. Because of natural herbal drugs are safe and effective; now herbal medicine and natural products are in big demand all over the World.

This book is an effort to put together the vast variety of medicinal plants found in India. The plants selected in the present book are fairly widely used in India for millennia. This book well helps in generating a global interest in *Ayurveda* and medicinal plants in India.

I hope that this compilation of most commonly used medicinal plants will be of use to Ayurvedic practitioners, students, Ethnobotanists and researchers as well as to medical and scientific communities

Dinesh Jadhav
'Kanchan',
30, Rajiv (Kasturba) Nagar,
Ratlam 457 001 (M.P.)
E-mail: dinj2ad@yahoo.com

ACKNOWLEDGEMENTS

First and foremost I express my deep sense of gratitude and indebtedness to Respected Dr. Shri Anil K. Goel, Deputy Director, Botanic Garden, National Botanical Research institute (NBRI), Lucknow (Uttar Pradesh) and Respected Shri Dr. V.P. Singh, Retd. Prof. & Head, IEMPS, Vikram University, Ujjain (M.P.) for encouragement, keen interest, support and co-operation.

I express my deep sense of gratitude to Respected Shri R.L. Chauhan, Retd. Professor & Head, Department of English, Govt. Arts and Science College, Ratlam (M.P.) and Respected Dr. (Mrs.) Bhavana Deshpandey, Professor, Department of Mathematics, Govt. Arts and Science College, Ratlam (M.P.) for encouragement, keen interest, support and valuable co-operation.

It is also a pleasure to thank Mr. Manish Solanki, Library incharge, Central Library, Pt. Shivshakti Lal Sharma Ayurvedic Medical College and Hospital, Ratlam (M.P.), Mr. Nilesh Shukla, Lab. Technician, Govt. Arts and Science College, Ratlam (M.P.), Mr. Shirish Mehra, Examination In-charge, Govt. Arts and Science College, Ratlam (M.P.), Mr. Narendra Singh Parihar (Johny), Computer Operator, Ratlam (M.P.) and Mr. Deepak Parmar (Deepu), Computer Operator, Govt. Arts and Science College, Ratlam (M.P.) for keen interest, support and co-operation.

My heartfelt thanks go to Respected Shri Praveen Bohra, Consulting Engineer & Architect, Bohra and Associates, Ratlam for his moral support and co-operation.

Lastly, I take this opportunity to express my gratitude to Respected Shri Pawan Kumar Ji, Managing Director, Scientific Publishers, Jodhpur, for accepting the Manuscript and also for guidance and helpful suggestions. Without his encouragement and support this book would not have been completed.

Dinesh Jadhav

ABBREVIATIONS USED IN THE BOOK

A.ma.	:	Abhidhanamanjari
A. Sam.	:	Astangasamgraham
A. vi.	:	Ayurveda vigyan
Bha.pra.	:	Bhavprakash
Bri.ma.	:	Brind Madhav
Ca. ka.	:	Carakasamhita kalpasthanam
Ca. Sam	:	Carakasamhita.
Ca.su.	:	Carakasamhita Sutrasthanam
Dha.ni.	:	Dhanvantarinighantu
Dra.gu.sam.	:	Dravyagunasamgraham
Dra.vi.	:	Dravyagunavijnan
Hr. Pri.	:	Hrdayapriya
Kai.ni.	:	Kaiyadevanighantu
Ma. ni.	:	Madanadinighantu
Ma. Vi.	:	Madanavinodam
Ni.ra.	:	Nighanturatnakaram
Px	:	Part not specified
Ra.ni.	:	Rajanighantu
Ra. va.	:	Rajavallabham
Sa. ni.	:	Saligramanighantu
Sa.ni.bhu.	:	Saligramanighantubhusanam
Si. Bhe. Ma.	:	Sidhabhesajamanimala
So.ni.	:	Sodhalanighantu
Su. Su.	:	Susrutasamahita Sutrasthanam
Sva.	:	Svayamkriti
Syn.	:	Synonyms

KEY TO TRANSLITERATION

Sanskrit	Transliteration key	Sanskrit	Transliteration key
अ	A, a	ट	ṭa
आ	Ā , ā	ठ	ṭha
इ	I, i	ड	ḍa
ई	Ī, ī	ढ	ḍha
उ	U, u	ण	ṇa
ऊ	Ū, ū	त	ta
ऋ	Ṛ, ṛ	थ	tha
–	Ṝ, ṝ	द	da
ए	E, e	ध	dha
–	Ē, ē	न	na
ऐ	Ai, ai	प	pa
ओ	O, o	फ	pha
–	Ō, ō	ब	ba
औ	Ou/ ou	भ	bha
अं	aṃ/ ama	म	ma
अः	aḥ	य	ya
क	ka	र	ra
ख	kha	ल	la
ग	ga	व	va
घ	gha	ष	ṣa
ड़.	naa	श	śa
च	ca	स	sa
छ	cha	ह	ha
ज	ja	ळ	ḷa
झ	jha	क्ष	kṣa
ञ	ña	ज्ञ	gna

INTRODUCTION

Man has depended on plants from time immemorial; plants have been a rich source for therapeutics. Medicinal plants belong to the earliest known health care products that have been used by the mankind. Over three-quarters of the World population rely on the use of traditional medicines for their primary health care needs. The medicinal plants are not only major components of the many formulations used in indigenous systems of medicine but also of a large number of drugs in Allopathy. In India, with one of the oldest ethnobotanical tradition in the World, medicinal plants are the backbone of all major systems of medicine. One of the earliest references to the Medicinal properties of plants is in the *Rigveda* (3500-1800 BC). *Ayurveda,* the ancient Indian system of medicine based on that medicinal property of plants. There are many more useful medicinal plants has been described under Ayurvedic medicinal system by several authors at that time i.e. '*Charakh*' and '*Susruta*' etc. Their Ayurvedic text manifests that an incredible 1600 of the 2000 old Ayurvedic formulations are plant based. After that there are many more books have been written under heading medicinal plants of India, but those books could not fulfil all the requirements of research scholar and readers. The present book is focused on medicinal plants of India. This book is differing from other books on medicinal plants because in this book each taxon contains ethno-medicinal uses with their Ayurvedic medicinal value. The ethno-medicinal value of the taxa is updated upto 2006. The sources of information included survey of published literature including primary, secondary and tertiary sources related to medicinal plants. The description on each plant species includes regional names, Distribution, Taxonomic description, Chemical constituents, Ayurvedic Medicinal properties, medicinal uses and Ethnomedicinal uses. This book is helpful for ethnobotanists, Ayurvedic medical practitioner, students and researchers as well as other reader's interested in the field of ethno-medicine.

Arrangement - This book (Volume 3) deals with 200 plants species of Angiosperms. Species of plants are presented in Alphabetical order of their Botanical names. In dealing with each species, after its Botanical name, all available Synonyms are mentioned. This is followed by names popularly used in English and in other Indian languages with its distribution. To help to identify the plant species, the taxonomic description is given. Available information about the chemical constituents of each species is given. Sanskrit Shlokas from relevent Nighantus describing the Ayurvedic Medicinal properties are first given in Devanagary script followed by its rendering in Roman script using internationally recognized transliteration markings. This is followed by giving its action and uses according to Ayurvedic therapeutics. Information about its use in Ethnomedicinal practice is given, then.

CONTENTS
(VOLUME 1)

PLANTS COVERED

Abelmoschus moschatus (L.) Medic. (Malvaceae) 2
Abrus precatorius L. (Papilionaceae) 3
Abutilon indicum (L.) Sweet (Malvaceae) 5
Acacia arabica auct. (Mimosaceae) 6
Acacia catechu Willd. (Mimosaceae) 8
Acalypha indica L. (Euphorbiaceae) 9
Achyranthes aspera L. (Amaranthaceae) 10
Adansonia digitata L. (Bombacaceae) 12
Adhatoda zeylanica Medic. (Acanthaceae) 13
Aegle marmelos (L.) Corr. ex Roxb. (Rutaceae) 15
Ageratum conyzoides L. (Asteraceae) 16
Ailanthus excelsa Roxb. (Simaroubaceae) 17
Alangium salvifolium (L.f.) Wang. (Alangiaceae) 19
Albizia lebbeck (L.) Willd. (Mimosaceae) 20
Allium cepa L. (Liliaceae) 22
Allium sativum L. (Liliaceae) 23
Aloe vera L. (Liliaceae) 24
Alstonia scholaris R.Br. (Apocynaceae) 25
Alternanthera sessilis (L.) R.Br. ex. DC. (Amaranthaceae) 26
Amaranthus spinosus L. (Amaranthaceae) 27
Amorphophallus campanulatus (Roxb.) Blume ex Decne (Aracaceae) 28
Anacardium occidentale L. (Anacardiaceae) 29
Anacyclus pyrethrum DC. (Asteraceae) 31
Andrographis paniculata (Burm. f.) Wall. ex Nees (Acanthaceae) 32
Annona squamosa L. (Annonaceae) 33
Anogeissus latifolia (Roxb. ex DC.) Wall. ex Guill. & Perr. (Combretaceae) 34
Anthocephalus cadamba (Roxb.) Miq. (Rubiaceae) 35
Argemone mexicana L. (Papaveraceae) 37
Argyreia speciosa Sweet. (Convolvulaceae) 38
Aristolochia indica L. (Aristolochiaceae) 39

Asparagus racemosus Willd. (Liliaceae) 41
Azadirachta indica A. Juss. (Meliaceae) 42
Bacopa monnieri (L.) Penn. (Scrophulariaceae) 44
Balanites aegyptiaca (L.) Del. (Simaroubaceae) 45
Barleria prionitis L. (Acanthaceae) 46
Bauhinia purpurea L. (Caesalpiniaceae) 47
Bauhinia variegata L. (Caesalpiniaceae) 49
Biophytum sensitivum (L.) DC. (Oxalidaceae) 50
Boerhavia diffusa L. (Nyctaginaceae) 51
Bombax ceiba L. (Bombacaceae) 52
Butea monosperma (Lamk.) Taub. (Papilionaceae) 54
Caesalpinia bonduc (L.) Taub. (Caesalpiniaceae) 56
Calotropis gigantea (L.) Ait. f. (Asclepiadaceae) 57
Calotropis procera (Ait.) R.Br. (Asclepiadaceace) 58
Cannabis sativa L.(Cannabinaceae) 60
Cassia fistula L. (Caesalpiniaceae) 61
Catharanthus roseus Don (Apocynaceae) 63
Cayratia carnosa (Wall. ex Wight) Gagnep. (Vitaceae) 64
Celastrus paniculatus Willd. (Celastraceae) 65
Celosia argentea L. (Amaranthaceae) 66
Centella asiatica (L.) Urban (Apiaceae) 67
Centratherum anthelminticum (Willd.) Kuntze (Asteraceae) 69
Chenopodium album L. (Chenopodiaceae) 70
Chlorophytum tuberosum Bak. (Liliaceae) 71
Cichorium intybus L. (Asteraceae) 72
Cissampelos pareira L. (Menispermaceae) 73
Cissus quadrangularis L. (Vitaceae) 74
Citrullus colocynthis (L.) Schrad. (Cucurbitaceae) 75
Cleome gynandra L. (Cleomaceae) 77
Cleome viscosa L. (Cleomaceae) 78
Clerodendrum inerme (L.) Gaertn. (Verbenaceae) 79
Clerodendrum phlomidis L. f. (Verbenaceae) 80
Clerodendrum serratum (L.) Moon. (Verbenaceae) 81
Clitoria ternatea L. (Papilionaceae) 82
Coccinia cordifolia Cogn. (Cucurbitaceae) 83
Cocculus hirsutus (L.) Diels. (Menispermaceae) 84
Coix lacryma - jobi L. (Poaceae) 86
Coleus aromaticus Benth. (Lamiaceae) 87
Colocasia esculenta (L.) Schott. (Aracaceae) 88
Commiphora mukul (Hook. ex Stocks) Engl. (Burseraceae) 89
Convolvulus pluricaulis Choisy. (Convolvulaceae) 90

Corida dichotoma Forst. f. (Boraginaceae) 91
Costus speciosus (Koenig ex Retz.) J.E. Smith (Zingiberaceae) 92
Crataeva magna (Lour.) DC. (Capparaceae) 94
Crinum asiaticum L. (Amaryllidaceae) 95
Crotalaria retusa L. (Papilionaceae) 96
Curculigo orchioides Gaertn. (Amaryllidaceae) 97
Curcuma amada Roxb. (Zingiberaceae) 98
Curuma longa L. (Zingiberaceae) 99
Cuscuta reflexa Roxb. (Convolvulaceae) 100
Cymbopogon martinii (Roxb.) Wats. (Poaceae) 101
Cynodon dactylon (L.) Pers. (Poaceae) 102
Cyperus rotundus L. (Cyperaceae) 103
Dalbergia sissoo Roxb.(Papilionaceae) 104
Datura metal L. (Solanaceae) 106
Datura stramonium L. (Solanaceae) 107
Desmodium gangeticum (L.) DC. (Papilionaceae) 108
Dichrostachys cinerea (L.) W. & A. (Mimosaceae) 109
Dioscorea alata L. (Dioscoreaceae) 110
Dioscorea bulbifera L. (Dioscoreaceae) 111
Diplocyclos palmatus (L.) Jeffrey (Cucurbitaceae) 112
Drypetes roxburghii (Wall.) Hurus. (Euphorbiaceae) 113
Echinops echinatus Roxb. (Asteraceae) 114
Eclipta prostrata (L.) L. (Asteraceae) 115
Embelia ribes Burm. f. (Myrsinaceae) 116
Embelia tsjeriam cottam (R. & S.) DC. (Myrsinaceae) 117
Enicostemma littorale Blume. (Gentianaceae) 118
Erythrina variegata L. (Papilionaceae) 119
Euphorbia antiquorum L. (Euphorbiaceae) 120
Euphorbia hirta L. (Euphorbiaceae) 121
Euphorbia neriifolia L. (Euphorbiaceae) 122
Euphorbia tirucalli L. (Euphorbiaceae) 124
Evolvulus alsinoides L. (Convolvulaceae) 125
Ficus benghalensis L. (Moraceae) 126
Ficus glomerata Roxb. (Moraceae) 127
Ficus religiosa L. (Moraceae) 128
Fumaria indica (Hassk.) Pugsley (Fumariaceae) 129
Gloriosa superba L. (Liliaceae) 130
Glycyrrhiza glabra L. (Papilionaceae) 132
Gmelina arborea Roxb. (Verbenaceae) 133
Grewia subinaequalis DC. (Tiliaceae) 134
Gymnema sylvestre (Retz.) R.Br. ex Sch. (Asclepiadaceae) 135

Haldina cordifolia (Roxb.) Ridsd. (Rubiaceae) 136
Helicteres isora L. (Sterculiaceae) 137
Hemidesmus indicus (L.) R.Br. (Asclepiadaceae) 138
Hibiscus rosa-sinensis L. (Malvaceae) 140
Holarrhena pubescens (Buch.- Ham.) Wall. ex G. Don. (Apocynaceae) 141
Holoptelea integrifolia (Roxb.) Planch. (Ulmaceae) 142
Hygrophila auriculata (Sch.) Heine (Acanthaceae) 143
Jatropha curcas L. (Euphorbiaceae) 144
Jatropha gossypifolia L. (Euphorbiaceae) 146
Kalanchoe pinnata (Lam.) Pers. (Crassulaceae) 147
Lantana camara L.(Verbenaceae) 148
Lawsonia inermis L. (Lythraceae) 149
Leucas cephalotes (Roth.) Spreng. (Lamiaceae) 150
Limonia acidissima L. (Rutaceae) 151
Luffa acutangula (L.) Roxb. var. *amara* (Roxb.) C.B. Clarke (Cucurbitaceae) 152
Madhuca indica Gmel. (Sapotaceae) 153
Mangifera indica L. (Anacardiaceae) 155
Melia azedarach L. (Meliaceae) 156
Mimosa pudica L. (Mimosaceae) 158
Mimusops elengi L. (Sapotaceae) 159
Mimusops hexandra (Roxb.) Dub. (Sapotaceae) 160
Mitragyna parvifolia (Roxb.) Korth. (Rubiaceae) 161
Momordica charantia L. (Cucurbitaceae) 162
Momordica dioica Roxb. ex Willd. (Cucurbitaceae) 164
Moringa oleifera Lam. (Moringaceae) 165
Morus alba L. (Moraceae) 167
Mucuna pruriens (L.) DC. (Papilionaceae) 168
Mukia madraspatana (L.) Roem. (Cucurbitaceae) 169
Murraya koenigii (L.) Spr. (Rutaceae) 170
Nelumbo nucifera Gaertn. (Nymphaeaceae) 171
Nerium indicum Mill. (Apocynaceae) 173
Nyctanthes arbor-tristis L. (Oleaceae) 174
Nymphaea nauchali Burm f. (Nymphaeaceae) 175
Ocimum basilicum L. (Lamiaceae) 176
Ocimum canum Sims. (Lamiaceae) 177
Ocimum gratissimum L. (Lamiaceae) 178
Ocimum sanctum L. (Lamiaceae) 179
Oroxylum indicum (L.) Vent. (Bignoniaceae) 181
Oxalis corniculata L. (Oxalidaceae) 182
Pandanus fascicularis Lamk. (Pandanaceae) 183
Papaver somniferum L. (Papaveraceae) 184

Pergularia daemia (Forsk.) Chiov. (Asclepiadaceae) 186
Phyla nodiflora L. (Verbenaceae) 187
Phyllanthus emblica L. (Euphorbiacae) 188
Phyllanthus niruri Auct. non. L. (Euphorbiaceae) 189
Physalis minima L. (Solanaceae) 191
Plumbago zeylanica L. (Plumbaginaceae) 192
Pongamia pinnata (L.) Pierre (Papilionaceae) 193
Portulaca oleracea L. (Portulacaceae) 194
Portulaca quadrifida L. (Portulacaceae) 195
Psidium guajava L. (Myrtaceae) 196
Psoralea corylifolia L. (Papilionaceae) 198
Pterocarpus marsupium Roxb. (Papilionaceae) 199
Rauvolfia serpentina (L.) Benth. ex Kurz. (Apocynaceae) 200
Ricinus communis L. (Euphorbiaceae) 201
Rubia cordifolia L.(Rubiaceae) 202
Santalum album L. (Santalaceae) 204
Saraca asoca (Roxb.) De Wilde. (Papilionaceae) 205
Senna absus L. (Roxb.) (Caesalpiniaceae) 206
Senna auriculata L. (Roxb.) (Caesalpiniaceae) 207
Senna occidentalis (L.) Link. (Caesalpiniaceae) 208
Senna tora L. (Roxb.) (Caesalpiniaceae) 210
Sesbania grandiflora (L.) Poir. (Papilionaceae) 211
Sida cordifolia L. (Malvaceae) 212
Sida rhombifolia L. (Malvaceae) 213
Solanum indicum L. (Solanaceae) 215
Solanum nigrum L. (Solanaceae) 216
Solanum surattense Burm. f. (Solanaceae) 217
Symplocos racemosa Roxb. (Symplocaceae) 219
Syzygium aromaticum (L.) Merr. & Perr. (Myrtaceae) 220
Syzygium cumini (L.) Skeels (Myrtaceae) 221
Tamarindus indica L. (Caesalpiniaceae) 222
Tectona grandis L. f. (Verbenaceae) 224
Tephrosia purpurea Pers. (Papilionaceae) 225
Terminalia arjuna (Roxb.) Wt. & Arn. (Combretaceae) 226
Terminalia bellerica (Gaertn.) Roxb. (Combretaceae) 227
Terminellia chebula (Gaertn.) Retz. (Combritaceae) 228
Tinospora cordifolia (Willd.) Miers. ex Hk. f. & Th. (Menispermaceae) 229
Toona ciliata Roem. (Meliaceae) 231
Tribulus terrestris L. (Zygophyllaceae) 232
Urginea indica (Roxb.) Kunth. (Liliaceae) 233
Vernonia cinerea (L.) Less (Asteraceae) 234

Vitex negundo L. (Verbenaceae) 235
Withania somnifera (L.) Dunal (Solanaceae) 236
Woodfordia fruticosa (L.) Kurz. (Lythraceae) 238
Xanthium strumarium L. (Asteraceae) 239
Zingiber officinale Rosc. (Zingiberaceae) 240
Ziziphus mauritiana Lam. (Rhamnaceae) 241

CONTENTS
(VOLUME 2)

PLANTS COVERED

Abelmoschus esculentus (L.) Moench. (MALVACEAE) 3
Acacia leucophloea (Roxb.) Willd. (MIMOSACEAE) 4
Acacia sinuata (Lour.) Merr. (MIMOSACEAE) 5
Aconitum heterophyllum Wall. ex Royle (RANUNCULACEAE) 6
Aconitum napellus L. (RANUNCULACEAE) 7
Acorus calamus L. (ARACEAE) 8
Adenanthera pavonina L.(MIMOSACEAE) 10
Aerva lanata (L.) Juss. ex Schultes (AMARANTHACEAE) 11
Albizia odoratissima (L. f.) Benth. (MIMOSACEAE) 12
Allophylus serratus (Roxb.) Kurz (SAPINDACEAE) 13
Alpinia galanga (L.) Willd. (ZINGIBERACEAE) 14
Alstonia venenata R. Br. (APOCYNACEAE) 15
Ammania baccifera L. (LYTHRACEAE) 16
Amomum subulatum Roxb.(ZINGIBERACEAE) 17
Anamirta cocculus (L.) Wight & Arn. (MENISPERMACEAE) 18
Ananas comosus (L.) Merrill (BROMELIACEAE) 19
Anethum graveolens L. (APIACEAE) 20
Anisomeles malabarica (L.) R. Br. (LAMIACEAE) 21
Annona reticulata L. (ANNONACEAE) 22
Aphanamixis polystachya (Wall.) Parker (MELIACEAE) 23
Aquilaria agallocha Roxb.(THYMELEACEAE) 24
Arachis hypogaea L.(PAPILIONACEAE) 25
Areca catechu L. (ARECACEAE) 26
Aristolachia bracteolata Lam. (ARISTOLOCHIACEAE) 27
Artemisia nilagirica (Clarke) Pamp. (ASTERACEAE) 28
Artocarpus heterophyllus Lam. (MORACEAE) 29
Asclepias curassavica L. (ASCLEPIADACEAE) 31

Asparagus adscendens Roxb. (LILIACEAE) 32
Averrhoa carambola L. (OXALIDACEAE) 33
Baliospermum montanum (Willd.) Muell-Arg. (EUPHORBIACEAE) 34
Bambusa arundinacea (Retz.) Willd. (BOMBUSACEAE) 35
Barringtonia acutangula (L.) Gaertn. (BARRINGTONIACEAE) 36
Basella alba L. var. *rubra* (L.) Stewart (BASELLACEAE) 38
Bauhinia racemosa Lam. (CAESALPINIACEAE) 39
Benincasa hispida (Thunb.) Cogn. (CUCURBITACEAE) 40
Berberis aristata DC. (BERBERIDACEAE) 41
Beta vulgaris L. (CHENOPODIACEAE) 42
Betula utilis D. Don. (BETULACEAE) 43
Bixa orellana L. (BIXACEAE) 44
Blumea lacera (Burm. f.) DC. (ASTERACEAE) 45
Borassus flabellifer L. (ARECACEAE) 46
Boswellia serrata Roxb. ex Colebr. (BURSERACEAE) 47
Brassica juncea Czern. & Coss. m(BRASSICACEAE) 48
Buchanania lanzan Spreng. (ANACARDIACEAE) 49
Caesalpinia sappan L. (CAESALPINIACEAE) 51
Callicarpa macrophylla Vahl (VERBENACEAE) 52
Calophyllum inophyllum L. (CLUSIACEAE) 53
Calycopteris floribunda Lam. (COMBRETACEAE) 54
Camellia sinensis (L.) O. Kuntze (THEACEAE) 55
Capparis decidua (Forsk.) Edgew. (CAPPARACEAE) 55
Capsicum annuum L. (SOLANACEAE) 57
Cardiospermum halicacabum L. (SAPINDACEAE) 58
Careya arborea Roxb. (BARRINGTONIACEAE) 59
Carica papaya L. (CARICACEAE) 60
Carthamus tinctorius L. (ASTERACEAE) 62
Caryota urens L. (ARECACEAE) 63
Ceiba pentandra (L.) Gaertn. (BOMBACACEAE) 64
Cicer arietinum L. (PAPILIONACEAE) 65
Cinchona officinalis L. (RUBIACEAE) 66
Cinnamomum camphora (L.) Presl (LAURACEAE) 67
Cinnamomum tamala (Buch.-Ham.) Nees. (LAURACEAE) 68
Cinnamomum verum L. Presl (LAURACEAE) 69
Citrus limon (L.) Burm. f. (RUTACEAE) 70
Citrus medica L. (RUTACEAE) 71

Citrus reticulata Blanco (RUTACEAE)	72
Cochlospermum religiosum (L.) Alston. (COCHLOSPERMACEAE)	73
Cocus nucifera L. (ARECACEAE)	74
Coffea arabica L. (RUBIACEAE)	75
Coptis teeta Wall. (RANUNCULACEAE)	76
Coriandrum sativum L. (APIACEAE)	77
Cressa critica L. (CONVOLVULACEAE)	78
Crocus sativus L. (IRIDACEAE)	79
Croton tiglium L. (EUPHORBIACEAE)	80
Cuminum cyminum L. (APIACEAE)	81
Curcuma aromatica Salisb. (ZINGIBERACEAE)	82
Cymbopogon citratus (DC.) Stapf (POACEAE)	83
Cyperus esculentus L. (CYPERACEAE)	84
Daucus carota L. (APIACEAE)	84
Desmostachys bipinnata (L.) Stapf (POACEAE)	85
Dillenia indica L. (DILLENIACEAE)	86
Diospyros malabarica (Desr.) Kostel. (EBENACEAE)	87
Elaeocarpus sphaericus (Gaertn.) K. Schum. (ELAEOCARPACEAE)	88
Elettaria cardamomum (L.) Maton. (ZINGIBERACEAE)	89
Eleusine coracana (L.) Gaertn. (POACEAE)	90
Eucalyptus globulus Labill. (MYRTACEAE)	91
Eupatorium triplinerve Vahl. (ASTERACEAE)	92
Euryalc fcrox Salsb. (NYMPHAEACEAE)	93
Ferula asafoetida L. (APIACEAE)	94
Ficus hispida L. f. (MORACEAE)	95
Ficus microcarpa L. f. (MARACEAE)	96
Flacourtia indica (Burm. f.) Merr. (FLACOURTIACEAE)	96
Foeniculum vulgare Mill. (APIACEAE)	98
Fritillaria roylei Hook. f. (LILIACEAE)	99
Garcinia gummi-gutta (L.) Robs. (CLUSIACEAE)	99
Garcinia morella (Gaertn.) Desr. (CLUSIACEAE)	100
Gardenia gummifera L. f. (RUBIACEAE)	101
Glycomis arborea (Roxb.) DC. (RUTACEAE)	102
Gossypium arboreum L. (MALVACEAE)	103
Gossypium herbaceum L. (MALVACEAE)	104
Helianthus annus L. (ASTERACEAE)	105
Hiptage benghalensis (L.) Kurz (MALPIGHIACEAE)	106

Hordeum vulgare L. (POACEAE)	108
Hydnocarpus laurifolia (Dennst.) Sleumer (FLACOURTIACEAE)	109
Hyoscyamus niger L. (SOLANACEAE)	110
Ichnocarpus frutescens (L.) R. Br. (APOCYNACEAE)	111
Indigofera tinctoria L. (PAPILIONACEAE)	112
Inula racemosa Hook. f. (ASTERACEAE)	113
Ipomoea batatas (Linn.) Lam. (CONVOLVULACEAE)	114
Ipomoea mauritiana Jacq. (CONVOLVULACEAE)	115
Ipomoea nil (Linn.) Roth. (CONVOLVULACEAE)	116
Jasminum grandiflorum L. (OLEACEAE)	117
Jasminm sambac (Linn.) Ait. (OLEACEAE)	118
Juglans regia L. (JUGLANDACEAE)	119
Justicia gendarussa Burm. f. (ACANTHACEAE)	120
Kaempferia rotunda L. (ZINGIBERACEAE)	121
Kyllinga nemoralis (Forster) Dandy. ex Hutch.(CYPERACEAE)	122
Lablab purpureus L.(PAPILIONACEAE)	123
Lagenaria siceraria (Mol.) Standley. (CUCURBITACEAE)	124
Lens culinaris Med.(PAPILIONACEAE)	125
Lepidium sativum L. (BRASSICACEAE)	126
Linum usitatissimum L. (LINACEAE)	127
Lodoicea maldivica (Poir.) Pers. (ARECACEAE)	128
Loranthus longiflorus L. var. *falcatus* (L. f.) Kurz. (LORANTHACEAE)	129
Luffa cylindrica (L.) M. Roem.(CUCURBITACEAE)	130
Lycopersicon esculentum Mill. (SOLANACEAE)	131
Malaxis acuminata D. Don.(ORCHIDACEAE)	132
Mallotus philippensis (Lam.) Muell.-Arg. (EUPHORBIACEAE)	133
Mentha arvensis L.(LAMIACEAE)	134
Mentha piperita L. (LAMIACEAE)	135
Merremia emarginata (Burm. f.) Hall. f. (CONVOLVULACEAE)	135
Mesua nagassarium (Burm. f.) Kosterm.(CLUSIACEAE)	136
Michelia champaca L. (MAGNOLIACEAE)	137
Mirabilis jalapa L. (NYCTAGINACEAE)	138
Murraya paniculata (L.) Jack. (RUTACEAE)	139
Musa paradisiaca L. (MUSACEAE)	140
Myrica nagi Thunb. (MYRICACEAE)	141
Myristica fragrans Houtt.(MYRISTICACEAE)	142
Myristica malabarica Lam. (MYRISTICACEAE)	143

Nardostachys jatamansi DC. (VALERIANACEAE)	144
Naregamia alata Wight & Arn. (MELIACEAE)	145
Neptunia oleracea Lour. (MIMOSACEAE)	146
Nicotiana tabacum L. (SOLANACEAE)	146
Nigella sativa L. (RANUNCULACEAE)	147
Oldenlandia corymbosa L. (RUBIACEAE)	148
Operculina turpethum (L.) Silva Manso (CONVOLVULACEAE)	149
Pedalium murex L. (PEDALIACEAE)	150
Phaseolus trilobatus (L.) Schreb. (PAPILIONACEAE)	151
Phoenix dactylifera L. (ARECACEAE)	152
Phoenix sylvestris Roxb. (ARECACEAE)	153
Picrorhiza kurroa Auct. non Royle (SCROPHULARIACEAE)	154
Piper betle L. (PIPERACEAE)	155
Piper cubeba L.f. (PIPERACEAE)	156
Piper longum L. (PIPERACEAE)	157
Piper nigrum L. (PIPERACEAE)	158
Piper sylvaticum Roxb. (PIPERACEAE)	159
Pistacia chinensis Bunge (ANACARDIACEAE)	160
Pisum sativum L. (PAPILIONACEAE)	161
Plantago ovata Forssk. (PLANTAGINACEAE)	162
Pluchea lanceolata Oliver & Hiern. (ASTERACEAE)	163
Plumeria rubra L. (APOCYNACEAE)	164
Premna corymbosa Rottl. (VERBENACEAE)	165
Prunus amygdalus Batsch. (ROSACEAE)	166
Prunus cerasoides D. Don (ROSACEAE)	167
Prunus domestica L. (ROSACEAE)	168
Prunus persica Benth. & Hook. f. (ROSACEAE)	169
Pterocarpus santalinus L.f. (PAPILIONACEAE)	170
Pueraria tuberosa DC. (PAPILIONACEAE)	171
Punica granatum L. (PUNICACEAE)	172
Pyrus malus L. (ROSACEAE)	173
Quisqualis indica L. (COMBRETACEAE)	174
Randia dumetorum Poir. (RUBIACEAE)	175
Raphanus sativus L. (BRASSICACEAE)	176
Rhinacanthus nasuta (L.) Kurz. (ACANTHACEAE)	177
Rosa damascena Mill. (ROSACEAE)	178
Sansevieria roxburghiana Schult. f. (AGAVACEAE)	178

Saccharum officinarum L. (POACEAE) 179
Shorea robusta Gaertn. f. (DIPTEROCARPACEAE) 180
Smilax china L. (LILIACEAE) 182
Solanum melongena L. (SOLANACEAE) 183
Soymida febrifuga A. Juss. (MELIACEAE) 184
Spinacea oleracea L. (CHENOPODIACEAE) 185
Strychnos nux-vomica L. (STRYCHNACEAE) 186
Swertia chirayita (Roxb. ex Flem.) Kar. (GENTIANACEAE) 187
Tabernaemontana divaricata (L.) R. Br. (APOCYNACEAE) 188
Tagetus erecta (L.) willd. (ASTERACEAE) 190
Teramnus labilis (L. F.) Spr. (PAPILIONACEAE) 191
Terminalia crenulata Roth.(COMBRETACEAE) 192
Thespesia populnea (L.) Soland. ex Correa. (MALVACEAE) 193
Thevetia peruviana (Pers.) Schum. (APOCYNACEAE) 194
Trapa bispinosa (Roxb.) Makino. (TRAPACEAE) 195
Trichosanthes dioica Roxb. (CUCURBITACEAE) 196
Trigonella foenum-graceum L. (PAPILIONACEAE) 197
Tylophora indica (Burm. f.) Merr. (ASCLEPIADACEAE) 198
Uraria picta Desv. (PAPILIONACEAE) 199
Vetiveria zizanioides (L.) Nash. (POACEAE) 200
Viola odorata L. (VIOLACEAE) 201
Vitis vinifera Dunal. (VITACEAE) 202
Wrightia tinctoria (Roxb.) R. Br. (APOCYNACEAE) 203
Zanthoxylum armatum DC. (RUTACEAE) 204
Zea mays L. (Poaceae) 205

CONTENTS

Preface *vii*

Acknowledgements *ix*

Abbreviations *xi*

Key to Transliteration *xiii*

Introduction *xv*

Contents of Volume - 1 (Plants covered) *xvii*

Contents of Volume - 2 (Plants covered) *xxiii*

MEDICINAL PLANTS OF INDIA

1. *Abroma angusta* L.f. (STERCULIACEAE) 1
2. *Acacia farnesiana* Willd. (MIMOSACEAE) 2
3. *Acacia pennata* (L.) Willd. (MIMOSACEAE) 3
4. *Acacia polyantha* Willd. (MIMOSACEAE) 4
5. *Achillea millefolium* L. (ASTERACEAE) 5
6. *Albizia procera* Benth. (MIMOSACEAE) 6
7. *Alhagi camelorum* Fisch. (FABACEAE) 7
8. *Alocasia indica* (Roxb.) Schott. (ARACEAE) 8
9. *Althaea officinalis* L. (MALVACEAE) 9
10. *Altingia excelsa* Noronha. (HAMAMELIDACEAE) 10
11. *Amorphophallus sylvaticus* (Roxb.) Kunth. (ARACEAE) 10
12. *Ampelocissus latifolia* (Roxb.) Planch. (VITACEAE) 11
13. *Angelica glauca* Edgew. (APIACEAE) 12
14. *Anisomeles indica* (L.) Kuntze. (LAMIACEAE) 13
15. *Artemisia maritima* L. (ASTERACEAE) 14
16. *Artocarpus communis* Forst. (MORACEAE) 15
17. *Artocarpus hirsutus* Lam.(MORACEAE) 15
18. *Arundo donax* L. (POACEAE) 16
19. *Atalantia monophylla* Corr.(RUTACEAE) 17
20. *Brassica nigra* (L.) Koch. (BRASSICACEAE) 18

21. *Caesalpinia digyna* Rottl. (CAESALPINIACEAE) 19
22. *Cajanus cajan* (L.) Millsp. (FABACEAE) 19
23. *Calamus rotang* L. (ARECACEAE) 20
24. *Canavalia gladiata* (Jacq.) DC. (FABACEAE) 21
25. *Canscora decussata* (Roxb.) Schult. (GENTIANACEAE) 22
26. *Canthium parviflorum* Lam. (RUBIACEAE) 23
27. *Capparis zeylanica* L. (CAPPARACEAE) 24
28. *Carissa carandas* L. (APOCYNACEAE) 25
29. *Cassia alata* L. (CAESALPINNIACEAE) 26
30. *Cassia senna* L. (CAESALPINNIACEAE) 27
31. *Cerbera odollam* Gaertn. (APOCYNACEAE) 28
32. *Chenopodium anthelminticum* Bert. (CHENOPODIACEAE) 29
33. *Chonemorpha fragrans* (Moon.) Alston. (APOCYNACEAE) 30
34. *Chrozophora rottleri* (Geis.) A. Juss. ex Spreng. (EUPHORBIACEAE) 31
35. *Cicca distichia* L. (EUPHORBIACEAE) 32
36. *Citrus aurantifolia* (Christm.) Swingle (RUTACEAE) 33
37. *Clematis triloba* Heyne ex Roth. (RANUNCULACEAE) 34
38. *Colchicum luteum* Baker. (LILIACEAE) 34
39. *Coleus vettiveroides* K.C. Jacob. (LAMIACEAE) 35
40. *Commelina benghalensis* L. (COMMELINACEAE) 36
41. *Commiphora myrrha* (Nees.) Engl. (BURSERACEAE) 36
42. *Convolvulus arvensis* L. (CONVOLVULACEAE) 37
43. *Corchorus olitorius* L. (TILIACEAE) 38
44. *Coscinium fenestratum* (Gaertn.) Colebr. (MENISPERMACEAE) 39
45. *Crinum defixum* Ker.-Gawal. (AMARYLLIDACEAE) 39
46. *Croton oblongifolius* Roxb. (EUPHORBIACEAE) 40
47. *Cucumis melo* var. utilissimus (Roxb.) Duthie & Fuller. (CUCURBITACEAE) 41
48. *Cucumis sativus* L. (CUCURBITACEAE) 42
49. *Cucumis trigonus* Roxb. (CUCURBITACEAE) 43
50. *Cucurbita maxima* Duchesne ex Lam. (CUCURBITACEAE) 44
51. *Cyamopsis tetragonoloba* (L.) Taub. (FABACEAE) 45
52. *Cyathula prostrata* (L.) Blume. (AMARANTHACEAE) 46
53. *Cyclea peltata* (Lam.) Hook. F. & Thoms. (MENISPERMACEAE) 47
54. *Cyperus scariosus* R.Br. (CYPERACEAE) 48
55. *Delphinium denudatum* Wall. ex Hook. f. & Thoms. (RANUNCULACEAE) 49
56. *Dendrocalamus stirctus* (Roxb.) Nees. (POACEAE) 50

57. *Desmodium triflorum* (L.) DC. (FABACEAE) 50
58. *Digera arvensis* (L.) DC.(FABACEAE) 51
59. *Digitalis purpurea* L. (SCROPHULARIACEAE) 52
60. *Diospyros melanoxylon* Roxb. (EBENACEAE) 52
61. *Diospyros malabarica* Kostal. (EBENACEAE) 53
62. *Emilia sonchifolia* (L.) DC.(ASTERACEAE) 54
63. *Euphorbia thymifolia* L. (EUPHORBIACEAE) 55
64. *Ferula narthex* Boiss.(APIACEAE) 56
65. *Ficus arnottiana* (Miq.) Miq. (MORACEAE) 57
66. *Flacourtia jangomas* Rausch. (FLACOURTIACEAE) 57
67. *Garuga pinnata* Roxb. (BURSERACEAE) 58
68. *Gentiana kurroo* Royle. (GENTIANACEAE) 59
69. *Grewia tilifolia* Vahl.(TILIACEAE) 60
70. *Habenaria edgeworthii* Hook. f. ex. Collett. (ORCHIDACEAE) 61
71. *Habenaria intermedia* D. Don.(ORCHIDACEAE) 62
72. *Hedyotis herbacea* L.(RUBIACEAE) 62
73. *Hedychium spicatum* Buch.-Ham. ex Smith (ZINGIBERACEAE) 63
74. *Heliotropium indicum* L.(BORAGINACEAE) 64
75. *Helleborus niger* L. (RANUNCULACEAE) 65
76. *Hibiscus aculeatus* Roxb.(MALVACEAE) 66
77. *Hibiscus cannabinus* L.(MALVACEAE) 66
78. *Holostemma ada-kodien* Schultes. (ASCLEPIADACEAE) 67
79. *Homonoia riparia* Lour.(EUPHORBIACEAE) 68
80. *Hugonia mystax* L. (LINACEAE) 69
81. *Hygroryza aristata* (Retz.) Nees ex Wight & Arn.(POACEAE) 70
82. *Illicium verum* Hook. f. (MAGNOLIACEAE) 70
83. *Ipomoea pes-caprae* (L.) R. Br. (CONVOLVULACEAE) 71
84. *Ipomoea sepiaria* Roxb. (CONVOLVULACEAE) 72
85. *Ixora coccinea* L. (RUBIACEAE) 73
86. *Jasminum angustifolium* (L.) Willd (OLEACEAE) 74
87. *Jasminum arborescens* Roxb. (OLEACEAE) 74
88. *Jasminum auriculatum* Vahl. (OLEACEAE) 75
89. *Jasminum multiflorum* (Burm. f.) Andr. (OLEACEAE) 76
90. *Justicia beddomei* (Clarke.) Bennet (ACANTHACEAE) 77
91. *Kaempferia galanga* L. (ZINGIBERACEAE) 77
92. *Lagerstroemia speciosa* (L.) Pers. (LYTHRACEAE) 78
93. *Launea coromandelica* (Houtt.) Merr. (ANACARDIACEAE) 79

94. *Leea indica* (Burm. f.) Merr. (LEEACEAE) 80
95. *Leptadenia reticulata* Wight & Arn. (ASCLEPIADACEAE) 81
96. *Lilium polyphyllum* D. Don (LILIACEAE) 81
97. *Limnophilla aromatica* (L.) Merr. (SCROPHULARIACEAE) 82
98. *Liquidamber orientalis* Mill. (HAMAMELIDACEAE) 83
99. *Macrotyloma uniflorum* (Lam.) Verdc. (FABACEAE) 84
100. *Malaxis muscifera* (Lindley) Kuntze. (ORCHIDACEAE) 85
101. *Manihot esculenta* Crantz. (EUPHORBIACEAE) 85
102. *Maranta arundinacea* L.(MARANTACEAE) 86
103. *Martynia annua* L. (MARTYNIACEAE) 87
104. *Mentha spicata* L. (LAMIACEAE) 88
105. *Merremia tridentata* (L.) Hallier f ssp. *Tridentata* (CONVOLVULACEAE) 89
106. *Monochoria vaginalis* Presl (PONTEDERIACEAE) 89
107. *Morinda pubescens* J.E. Smith. (RUBIACEAE) 90
108. *Mussaenda frondosa* L. (RUBIACEAE) 91
109. *Myxopyrum serratulum* A. W. Hill. (OLEACEAE) 92
110. *Naravelia zeylanica* (L.) DC. (RANUNCULACEAE) 93
111. *Nervilia aragoana* Gaud. (ORCHIDACEAE) 93
112. *Nilgirianthus ciliatus* (Nees.) Bremek. (ACANTHACEAE) 94
113. *Ocimum kilimandscharicum* Guerke. (LAMIACEAE) 95
114. *Ophiorrhiza mungos* L. (RUBIACEAE) 96
115. *Opuntia elatior* Mill. (CACTACEAE) 96
116. *Orchis latifolia* L. (ORCHIDACEAE) 97
117. *Origanum majorana* L. (LAMIACEAE) 98
118. *Oryza sativa* L. (POACEAE) 99
119. *Ougeinia oojeinensis* (Roxb.) Hochr. (FABACEAE) 100
120. *Oxystelma esculentum* R. Br. (ASCLEPIADACEAE) 101
121. *Panicum sumatrense* Roth. ex Roem. et Schult. (POACEAE) 102
122. *Paspalum scrobiculatum* L.(POACEAE) 102
123. *Pentapetes phoenicea* L. (STERCULIACEAE) 103
124. *Phaseolus mungo* (L.) Mant. (FABACEAE) 104
125. *Phoenix pusilla* Gaertn. (ARECACEAE) 105
126. *Phyllanthus reticulatus* Poir. (EUPHORBIACEAE) 105
127. *Piper brachystachyum* Wall. (PIPERACEAE) 106
128. *Plumbago indica* L. (PLUMBAGINACEAE) 107
129. *Podophyllum emodi* Wall. ex Hook. f. & Th.(BERBERIDACEAE) 108
130. *Polygonatum cirrhifolium* (Wall.) Royle (LILIACEAE) 109

131. *Polygonatum verticillatum* (L.) All. (LILIACEAE) 110
132. *Prunus armeniaca* L. (ROSACEAE) 111
133. *Pseudarthria viscida* (L.) Wight & Arn. (FABACEAE) 111
134. *Pterospermum canescens* Roxb. (STERCULIACEAE) 112
135. *Ranunculus sceleratus* L. (RANUNCULACEAE) 113
136. *Rhaphidophora laciniata* (Burm. f.) Merr. (ARACEAE) 114
137. *Rhinacanthus nasuta* (L.) Kurz.(ACANTHACEAE) 114
138. *Rhus parviflora* Roxb. (ANACARDIACEAE) 115
139. *Rotula aquatica* Lour. (BORAGINACEAE) 116
140. *Rungia repens* Nees. (ACANTHACEAE) 117
141. *Saccharum arundinaceum* Retz. (POACEAE) 117
142. *Saccharum munja* Roxb. (POACEAE) 118
143. *Saccharum spontaneum* L. (POACEAE) 119
144. *Salacia reticulata* Wight. (HIPPOCRATEACEAE) 120
145. *Salvadora persica* L. var. *Wightiana* [Planch ex Thw.] Verdc. (SALVADORACEAE) 120
146. *Samadera indica* Gaertn. (SIMAROUBACEAE) 121
147. *Sapindus laurifolia* Vahl. (SAPINDACEAE) 122
148. *Sarcostemma acidum* (Voigt) Voigt (ASCLEPIADACEAE) 123
149. *Sarcostigma kleinii* Wight & Arn. (ICACINACEAE) 124
150. *Sassurea lappa* C.B. Clarke. (ASTERACEAE) 125
151. *Schleichera oleosa* (Lour.) Oken. (SAPINDACEAE) 126
152. *Schrebera swietenioides* Roxb. (OLEACEAE) 127
153. *Scindapsus officinalis* (Roxb.) Schott. (ARACEAE) 128
154. *Securinega leucopyrus* (Willd.) Muell-Arg. (EUPHORBIACEAE) 129
155. *Semecarpus anacardium* L. f. (ANACARDIACEAE) 129
156. *Sesamum indicum* L. (PEDALIACEAE) 131
157. *Setaria italica* (L.) P. Beauv. (POACEAE) 132
158. *Sida spinosa* L. (MALVACEAE) 133
159. *Sisymbrium irio* L. (BRASSICACEAE) 134
160. *Solanum dulcamara* L. (SOLANACEAE) 134
161. *Solena amplexicaulis* (Lam.) Gandhi (CUCURBITACEAE) 135
162. *Sonchus arvensis* Auct. Plur. non L. (ASTERACEAE) 136
163. *Sorghum vulgare* Pers. (POACEAE) 137
164. *Spondias pinnata* (L. f.) Kurz. (ANACARDIACEAE) 138
165. *Sterculia foetida* L. (STERCULIACEAE) 139
166. *Sterospermum chelonoides* DC. (BIGNONIACEAE) 140

167. *Stereospermum colais* (Buch.- Ham. ex Dillw.) Mabberley (BIGNONIACEAE) 141
168. *Streblus asper* Lour. (MORACEAE) 142
169. *Strychnos potatorum* L. f. (LOGANIACEAE) 143
170. *Syzygium jambos* (L.) Alston. (MYRTACEAE) 144
171. *Tabernaemontana heyneana* Wall. (APOCYNACEAE) 145
172. *Taraxacum officinale* Weber. (ASTERACEAE) 145
173. *Terminalia paniculata* Roth. (COMBRITACEAE) 146
174. *Thalictrum foliolosum* DC. (RANUNCULACEAE) 147
175. *Tinospora malabarica* (Lam.) Miers. (MENISPERMINACEAE) 148
176. *Toddalia asiatica* (Lin.) Lam. (RUTACEAE) 149
177. *Trachyspermum ammi* (L.) Sprangae. (APIACEAE) 150
178. *Trachyspermum roxburghianum* (DC.) Craib (APIACEAE) 151
179. *Tragia involucrata* L. (EUPHORBIACEAE) 152
180. *Trianthema portulacastrum* L. (AZOIACEAE) 153
181. *Trichodesma indicum* (L.) R. Br. (BORAGINACEAE) 154
182. *Trichosanthes cucumerina* L. (CUCURBITACEAE) 155
183. *Trichosanthes lobata* Roxb. (CUCURBITACEAE) 156
184. *Trichosanthes tricuspidata* Lour. (CUCURBITACEAE) 156
185. *Tridax procumbens* L. (ASTERACEAE) 157
186. *Triticum aestivum* L. (POACEAE) 158
187. *Triumfetta rhomboidea* Jacq. (TILIACEAE) 159
188. *Typha elephantina* Roxb. (TYPHACEAE) 160
189. *Uvaria narum* (Dunal.) Wall. ex Wight. (ANNONACEAE) 161
190. *Valeriana jatamansi* Jones. (VALERIANACEAE) 162
191. *Vateria idica* L. (DIPTEROCARPACEAE) 163
192. *Ventilago maderaspatana* Gaertn. (RHAMNACEAE) 164
193. *Vigna pilosa* Baker. (FABACEAE) 165
194. *Vigna radiata* (L.) Wilczek (FABACEAE) 166
195. *Vigna radiata* (L.) Wilczek var. *Sublobata* (Roxb.) Verdc. (FABACEAE) 166
196. *Vigna unguiculata* (L.) Walp. ssp. *cylindrica* (L.) Eselt. (FABACEAE) 167
197. *Wedelia chinensis* (Osbeck.) Merrill. (ASTERACEAE) 168
198. *Zanonia indica* L. (CUCURBIACEAE) 169
199. *Ziziphus nummularia* (Burm. f.) Wt. & Arn. (RHAMNACEAE). 169
200. *Ziziphus oenoplia* (L.) Mill. (RHAMNACEAE) 170

GLOSSARY OF MEDICAL TERMS 172

BIBLIOGRAPHY 180

INDEX OF ENGLISH, HINDI & SANSKRIT NAMES OF PLANTS 207

ABROMA ANGUSTA L.f. (STERCULIACEAE)

Regional & Other Names : *Bengali* – Olatkambol; *English* – Devil's cotton; *Gujarati* - Ulatkambal; *Hindi* - Ulatkambal; *Kannada* - Bivvahatti; *Marathi* – Olatkambol; *Sanskrit* - Pivari; *Tamil* – Sivapputtutti.

Distribution: Throughout India.

Taxonomic Description: A large shrub or small tree. Leaves repand-denticulate. Flowers 5 cm. across, axillary. Capsules thrice as long as persistent calyx, glabrous.

Medicinal Parts: Leaves, Root, Rootbark, Stem.

Chemical Constituents: Root contains Abromine (betaine), Friedelin, Abromasterol-A, Choline, β-Sitosterol, Stigmasterol, Octacosanol, Friedelin.

Root bark contains α-amyrin, Maslinic acid, Vanillic acid, Caffeic acid, Rhamnose, Arabinose, xylose, Mannose, Galactose, Glucose, Galacturonic acid, Glucuronic acid.

Leaves contain Taraxerol, Taraxeryl acetate, β-Sitosterol, Lupeol.

Medicinal Properties:

पीवरी योषिनी सा स्यात् योनिव्यापद्विनाशिनी ।
रजोदोषप्रशमनी प्रदरार्शोनिवारिणी ।। (द्रव्यगुण)

Pīvarī yosinī sāṣyāṭ yonivyāpadvināśinī
Rajodoṣaśamanī pradarārśonivārinī. (Dṛavyaguṇa)

Root / Root bark- Abortifacient, Emmenogogue, Uterine tonic.

Stem / Leaves - Anti-inflammatory, Demulcent, Diuretic.

Medicinal Uses: Root / Root bark- Dysmenorrhoea, Menorrhoea, Menstrual disorders, Sterility.

Leaves - Diabetes, Gonorrhoea, Headache due to Sinusitis, Rheumatic pain of joints, Uterine disorders.

Ethno- Medicinal Uses: Root, Bark- Tonic (Gantait *et al.*, 2006).

Leaf - Diabetes, rheumatic pain (Gantait *et al.*, 2006).

ACACIA FARNESIANA Willd. (MIMOSACEAE)

Regional & Other Names : *Bengali* – Gaya Babul; *English* – Cassie flower; *Gujarati* – Gandhelo babul; *Hindi* – Vilayati kikar; *Malayalam* - Arivelam; *Marathi* – Kankar; *Sanskrit* – Irimeda; *Tamil* – Kadivel; *Telugu* – Kasturi tuma.

Distribution: Cosmopolitan in the tropics, often planted.

Taxonomic Description: Tree, grows upto 7-8 m. height covered with small thorns; Paripinnate compound leaf; Yellow coloured small flowers; Head inflorescence; fruit legume.

Medicinal Parts: Bark.

Chemical Constituents: Flowers are the source of Cassie Perfume. Pods contain tannin. Seeds contain a Fatty oil.

Medicinal Properties:

इरिमेदः कषायोष्णो मुखदन्तगदास्रजित् ।
हन्ति कण्डूविषश्लेष्मकृमिकुष्ठविषव्रणान् ।। (भा. प्र.)

Irimedaḥ kaṣayōsnō mukhdantagadāsrajiṭ
Hanti kanduvisaśleśmakrimikuṣtha visavrinān̤. (Bha. Pra.)

Bark - Acrid, Alexetric, Anthelmintic, Antidysenteric, Hot.

Medicinal Uses: Gum - Aphrodisiac, Sweet, Tonic.

Bark - Bronchitis, Diseases of Mouth & Blood, Erysipelas, Excessive haemorrhage, Inflammation, Itching, Leucoderma, Stomatitis, Ulcers.

Ethno- Medicinal Uses: Bark - Astringent (Singh & Pandey, 1980); Malaria, Stomach pain (Pal & Jain, 1989); Astringent, demulcent (Basak, 2006); toothache (Sinha *et al.,* 2006).

Root - Antifertility (Pal & Jain, 1989); Toothache (Sinha *et al.,* 2006).

Leaf - Eye complaints (Joshi, 1982; Ghosh & Sensarma, 1997); Gonorrhoea (Singh & Pandey, 1980); Malaria, gonorrhoea (Sinha *et al.,* 2006).

Inflorescence - Venereal diseases (Sebastian & Bhandari, 1984).

Dried gum- Diarrhoea (Siddiqui & Husain, 1991).

ACACIA PENNATA (L.) Willd. (MIMOSACEAE)

Syn.: *Mimosa pennata* L.

Regional & Other Names : *Bengali* – Kuchui; *Gujarati* – Kherval; *Hindi* - Aila; *Malayalam* - Kareencha; *Marathi* – Shembi; *Sanskrit*- Ari;; *Tamil* – Indu; *Telugu* – Karusikaya.

Distribution: Bengal, Bihar, Central, Western and South India.

Taxonomic Description: Large, scandent bush, with polished bark. Prickles slightly recurved. Pinnae 8-15 pairs, with glands between the upper ones; leaflets 40-50 pairs. Flowers creamy-white or pale-yellow. Pods stalked, linear-oblong, glabrous.

Medicinal Parts: Roots.

Chemical Constituents: Bark contains lupeol and □-Spinasterol and Stem contain β-Sitosterol.

Medicinal Properties :

अरिः कषायकटुका तिक्ता रक्तार्तिपित्तनुत् । (रा.नि.)

Ariḥ kasāyakatukā tiktā rakṭtāṛtipiṭtanuṭ. (Ra. Ni.)

Bark- Acrid, Bitter, Sour.

Medicinal Uses: Bark- Asthma, Biliousness, Bronchitis, Diseases of Blood.

Ethno- Medicinal Uses: Bark - Dandruff (Das *et al.,* 1983); Snake bite (Das *et al.,* 1983); Gum trouble, Gastro intestinal trouble (Sinha *et al.,* 2006).

Root - Digestive Complaints (Maheshwari *et al.*, 1986); Gum trouble, Gastro intestinal trouble (Sinha *et al.*, 2006).

Leaf - Headache (Jain, 1965); Bodyache (Maheshwari *et al.*, 1986; Jain, 1965); Post delivery cases (Gopal & Chandra, 2003).

ACACIA POLYANTHA Willd. (MIMOSACEAE)

Syn.: *Acacia suma* (Roxb.) Buch.-Ham.

Regional & Other Names : *Bengali* – Saikanta; *English* – White cutch; *Hindi* - Svetakhair; *Kannada* - Mugali; *Malayalam* - Venkarinnali; *Marathi* – Kamtiya; *Sanskrit* - Kadarah; *Tamil* - Kovil; *Telugu* - Tellatumma.

Distribution: Bengal, Bihar, Western peninsula.

Taxonomic Description: Medium sized erect trees, with bipinnately compound leaves and white flowered spikes inflorescence.

Medicinal Parts : Heartwood.

Chemical Constituents: Seed - Crude protein, Peritosan, Nitrogen free extract, Curde Fibre, Calcium, Phosphorus, Iron, Niacin, Ascorbic acid.

Seed oil - Fatty acids: Lauric, Myristic, Palmitic Oleic, Linoleic, Linolenic, Erucic, Lignoceric acids.

Heartwood - Quercetin, Hyperin 3', 7-Dimethyl ether.

Stem bark - Proanthocyanidin.

Medicinal Properties:

श्वेतस्तु खदिरस्तिक्तः कषायः कटुरूष्णकः ।
कण्डूतिभूतकुष्ठघ्नः कफवातव्रणापहः ।। (रा. नि.)

Śvetaștu khadiraștiķtaḥ kasāyaḥ katurūṣnakaḥ
Kaṇdūtibhūtakuṣthagḥnaḥ kaphavātavṛaṇāpahaḥ. (Rā. Ni.)

Heartwood - Acrid, Anthelmintic, Astringent, Bitter, Depurative, Revulsive, Thermogenic.

Medicinal Uses: Heartwood - Diabetes, Epilepsy, Helminthiasis, Leprosy, Leucoderma, Pruritus, Rheumatism, Skin diseases.

Ethno- Medicinal Uses: Heartwood - Blood purifier (Chandra & Pandey, 1985).

ACHILLEA MILLEFOLIUM L. (ASTERACEAE)

Regional & Other Names : *English* – Yarrow; *Gujarati* -Biranjasif; *Hindi* – Gandana; *Kannada* –Rooamari; *Sanskrit* – Svetadurva.

Distribution: Western Himalaya, from Kashmir to Kumaon, 6,000-9,000 ft. North Asia, Europe, North America.

Taxonomic Description: An erect pubescent herbs, with White or Pale Pink flowers and oblong, flattened, shining Achenes.

Medicinal Parts: All parts.

Chemical Constituents : Plant contains Alkaloid-Achilleine; Salicylic acid; β-Sitosterol and its acetate, Inositol, Dulcitol, Mannitol, Betain, Choline, Trigonekkine, Betonicine, Stachydrine; Essential oil; HCN-Glucoside Achillein.

Medicinal Properties-

श्वेतदूर्वातिशिशिरा मधुरा वान्तिपित्तजित् ।
आमातिसारिकासध्नी रूच्या दाहतृषापहा ।। (रा. नि.)

Śvetadụrvātiśiśirā madhurā vạntipịttajịt
Āmātisārikāsag̱hnī rūc̣hyā dāhaṭrsāpahā. (Ra. Ni.)

Antiseptic, Antispasmodic, Astringent, Diaphoretic, Emmenagogue, Mild aromatic, Stimulant, Styptic, Tonic, Vulnerary.

Medicinal Uses: Burns, Cold, Colic, Fever, Flatulence, Haemorrhage.

Ethno-Medicinal Uses: Whole plant - Cold (Gaur *et al.,* 1983; Srivastava *et al.,* 1981); Fever (Gupta *et al.,* 1980; Gaur *et al.,* 1983; Srivastava *et al.,* 1981); Gastric Complaint (Gupta *et al.,* 1980); Epilepsy, Hysteria, Toothache, Worms (Srivastava *et al.,* 1981); Piles, Ulcer (Gaur *et al.,* 1983); Skin infections (Ballabh & Chaurasia, 2006).

Leaves - Carminative (Ara & Naqshi, 1992); Inflammed ears & eyes (Kapur & Nanda, 1992).

Flowers - Carminative (Ara & Naqshi, 1992).

Root - Dysentery (Kapur & Nanda, 1992).

ALBIZIA PROCERA Benth. (MIMOSACEAE)

Regional & Other Names : *Bengali* – Koroi; *English* – White Sirish; *Hindi* – Safed Siris; *Kannada* – Bellate; *Malayalam* – Karun Thagara; *Marathi* – Kinhai; *Sanskrit* – Kinihi; *Tamil* – Kondavagei; *Telugu* – Tella Chinduga.

Distribution: Central & East Himalayas, Bihar, Western Peninsula, Burma.

Taxonomic Description: Trees, with grey-white bark, 2-6 pairs of pinnae and 6-15 pairs of ovate-oblong leaflets. Flowers pale-white in panicles. Pods brown.

Medicinal Parts : Leaves.

Chemical Constituents : Seed contain Saponins.

Seed oil contain Fatty acid.

Leaves contain Tannins, Saponins, Hentriacontane, Hexacosanol, □-Spinasterol, Carotenoid, Protein, Oxalate, Nitrate.

Gum contain D-galactose, L-arobinose, L-rhamnose.

Heartwood contain 4-Isoflavones; Biochanin-A, Formononetin, Genistein daidzein.

Medicinal Properties:

किणिही तुवरा तीक्ष्णा कटुका विनियच्छति ।
दोषत्रयविषश्वित्रब्रणग्रहशिरोगदान् ।। (कै. नि.)

Kiṇihī tuvarā tīkṣṇā katukā viniyacḥchati
Dosaṭray visaṣviṭravṛaṇgrah ṣhirōgdāṇ. (Kai. Ni.)

Leaves - Insecticide.

Medicinal Uses: Leaves - Ulcers.

Ethno- Medicinal Uses: Stem bark - Backache (Joshi *et al.,* 1980); Antipyretic (Sinha *et al.,* 2006).

Seed - Body pain (Sharma *et al.,* 1984).

ALHAGI CAMELORUM Fisch. (FABACEAE)

Syn.: *A. Pseudalhagi* (Bieb.) Desv.; *A. maurorum* Desv. ex Baker.

Regional & Other Names : *Bengali* – Javasa; *English* – Camel thorn; *Gujarati* – Jawaso; *Hindi* – Bharbharra; *Kannada* – Billiduruva; *Marathi* – Jawasa; *Sanskrit* – Yavasa; *Tamil* – Kanchori; *Telugu* – Girikarmika.

Distribution: Gujarat, Punjab, U.P., Rajasthan, Central & South India.

Taxonomic Description: Perennial shrub; Branches covered with spines; simple leaf; Pink flowers; Legume fruit.

Medicinal Parts: Leaves, Fruits, Seeds.

Chemical Constituents: Plant contains tannins, Flavonoids, Quercetin, β-Phenethylamine.

Medicinal Properties:

यासः स्वादुः सरस्तिक्तस्तूवरः शीतलो लघुः ।
कफमेदोमदभ्रान्तिपित्तासृक्कुष्ठकासजित् ।। (भा. प्र.)

Yāsaḥ svāduḥ sarastiktastuvaraḥ śitalo laghuḥ
Kaphamedomadbhrantipiṭtāsrukkusthakāsjiṭ. (Bha. Pra.)

Antibacterial, Antibilious, Anticancer, Antirheumatic, Antiseptic, Bitter, Diaphoretic, Diuretic, Laxative.

Medicinal Uses: Twigs - Piles, Cough.

Fruits, Flowers - Piles.

Root - Swellings, Abscesses.

Leaf oil - Rheumatic pain.

Leaves - Asthma, Bronchitis, Cough.

Ethno- Medicinal Uses: Leaf - Asthma (Shah *et al.,* 1981; Sebastian & Bhandari, 1984); Rheumatism (Tripathi *et al.,* 1996); Fever (Shah *et al.,* 1981).

Flowers- Haemorrhoids (Tripathi *et al.,* 1996).

Plant- Laxative, diuretic, expectorant (Tripathi *et al.,* 1996).

ALOCASIA INDICA (Roxb.) Schott. (ARACEAE)

Syn.: *A. macrorhiza* (L.) G. Don.; *A. odora* Koch.; *Colocasia macrorhiza* Schott.

Regional & Other Names : *Bengali* – Mankachu; *English* – Giant taro; *Hindi* – Mankanda; *Kannada* – Maanaka; *Malayalam* - Marambu; *Marathi* – Alu; *Sanskrit* – Manaka; *Tamil* – Merukankilangu; *Telugu* – Chara kanda.

Distribution: The plant is found in wild in forest, Mountains & is also cultivated in Assam, West Bengal, Maharashtra & South India.

Taxonomic Description: Herb, spadix inflorescence with light Yellow or Green coloured flowers. Berry fruits.

Medicinal Parts: Rhizome, Leaf.

Chemical Constituents: Rhizome contains - Sterols as β-Sitosterol, Campesterol, Cholesterol; Amino acids: Citric, Malic, Ascorbic, Succinic, Oxalic acid; Glucose, Fructose, Sucrose, Calcium, Oxalate, Alocasin.

Medicinal Properties:

मानकः स्यान्महापत्रः कथ्यन्ते तद्रुष्णा अथ ।
मानकः शोथहृच्छीतो रक्तपित्तहरोलघुः ।। (भा. प्र.)

Mānakaḥ syāṇmahāpatraḥ kathyaṇte tadrusnā ath
Mānakaḥ sōthhrcchītō raktapiṭtaharolaghuḥ. (Bha. Pra.)

Corm- Cooling, Digestive, Diuretic, Laxative, Nutritive, Rubifacient.

Leaves - Antitumour, Astringent, Rubifacient, Styptic.

Medicinal Uses: Corm- Arthritis, Chronic Constipation, Colic, Earache, Inflammation, Piles, Snake bite.

Leaf - Colic, Snake bite, Vomiting.

Ethno- Medicinal Uses: Petioles- Anaemia (Borthakur, 1996).

Corm-Wound (Kumar & Goel, 1998); Gout, rheumatism (De & Parikh, 1985).

ALTHAEA OFFICINALIS L. (MALVACEAE)

Regional & Other Names : *Bengali* – Gandhabhadulia; *English* – Marsh mallow; *Gujarati* - Khaira; *Hindi* – Gandhali; *Malayalam* - Talanili; *Marathi-* Gulkhair; *Sanskrit* – Prasarani, Khatmi; *Tamil* – Simaitutti; *Telugu* – Savirela.

Distribution: Wild in the Kashmir & Punjab.

Taxonomic Description: Perennial herbs; Leaves ovate with cordate base and unequal teeth, stipules are linear-subulate; flowers peduncled, rosy, in axillary clusters. Seed is solitary and ascending.

Medicinal Parts: Whole plant.

Chemical Constituents : Roots contain Mucilage, Starch, Pectin, Sugar, Fixed oil, Crystallin Substance: Althin or Khatmine.

Seed contains Drying oil.

Flowers contain Red dye.

Medicinal Properties:

खत्मी तू मधुरा स्निग्धा पिच्छिलाशीतला गुरूः।
वातपित्तहरा श्लेष्मसारणी मूत्रला सरा ।।

Khaṭmi tu madhurā ṣnigdhā picchilā śitalā guruḥ.
Vātpịtthạrā śleṣhmsārinī mutralā sarā.

Root / Seeds - Analgesic, Emollient.

Medicinal Uses: Root / Seeds – Cough, Earache, Inflammation of the breast, Lumbago.

Ethno- Medicinal Uses: Traditionally, a paste of its roots forms a popular remedy, which is used as an ointment.

ALTINGIA EXCELSA Noronha. (HAMAMELIDACEAE)

Regional & Other Names : *Gujarati* – Silaras; *Hindi* – Silaras; *Malayalam* - Rasamala; *Marathi* – Shilaras; *Sanskrit* – Silhaka; *Tamil* – Neriyurishippal; *Telugu* – Shilarasamu.

Distribution: Assam, Bhutan, Pegu, Mergui.

Taxonomic Description: Tree, simple leaves, head inflorescence, consists of Yellow coloured flowers, fruit capsule.

Medicinal Parts: Resin.

Chemical constituents: Benzaldehyde, cinnamic aldehyde.

Medicinal Properties:

सिल्हकः कटुकः स्वादुः स्निग्धोष्णः शुक्रकान्तिकृत् ।
वृष्यः कण्ठयः स्वेदकुष्ठज्वरदाहग्रहापहः ।। (भा. प्र.)

Silhakaḥ katukaḥ svaduḥ sṇigdhosnoḥ sukrakāṇtikruṭ
Vrisyaḥ kantyaḥ svedkusthajvardāhagrahāpahah. (Bha. Pra.)

Resin - Antipyretic, Antiscorbutic, Aphrodisiac, Aphrodisiac, Bitter, Carminative, Expectorant, Pungent, Stomachache.

Medicinal Uses: Resin -Biliousness, Bronchitis, Diseases of the bladder, Hydrocoel, Itching, Leucoderma, Lumbago, Menorrhagia, Pulmonary & Renal trouble, Scabies, Vesicular calculi.

AMORPHOPHALLUS SYLVATICUS (Roxb.) Kunth. (ARACEAE)

Syn.: *Synantherias sylvatica* (Roxb.) Schott.; *Amorphophallus zeylanicus* Blume.

Regional & Other Names: *English* - Wild suran; *Marathi* – Vajramuta; *Sanskrit* – Vajra kanda; *Tamil* - Kattukkarani; *Telugu* - Adavi kanda.

Distribution: Tamilnadu state in the Circars, Nilgiris & Coorg.

Taxonomic Description: Bulbiferous herb.

Medicinal Parts: Rhizome.

Chemical Constituents: Calcium oxalate.

Medicinal Properties -

लोहितः सूरणः कण्डूकर्ता विष्टम्भको लघुः ।
तुवरो रूक्षकटुको रूच्यो विशद्दीनः ।। (नि. र.)

Lohitaḥ sūraṇaḥ kaṇdūkaṛtā viṣtaṃbhako laghuḥ.
Tuvarō rūksakatuko rūcyo viśaddīpanaḥ. (Ni. Ra.)

Tuber - Anthelmintic, Aphrodisiac, Pungent, Stimulant of urinary tract.

Medicinal Uses: Seed - Pains, Piles, Gland enlargement, Toothache, Tumours.

Ethno- Medicinal Uses: Leaf - Diarrhoea (Thomas & Britto, 2003)

Root - Dysentery (Thomas & Britto, 2003).

AMPELOCISSUS LATIFOLIA (Roxb.) Planch. (VITACEAE)

Syn.: *Vitis latifolia* Roxb.

Regional & Other Names : *Bengali* – Govila; *English* – Jungle grape vine; *Gujarati* - Junglidrakh; *Hindi* – Panibel; *Malayalam* - Karantavalli; *Marathi* – Golinda; *Tamil* - Kattukkodimundirigui; *Telugu* – Bedasativa.

Distribution: Sub-Himalayan tract from the Sutlej eastwards to Kumaon up to 4,000 ft., Assam, Sylhet, Konkan, Western Ghats from Bombay to the Nilgiris & Anamalais, Deccan.

Taxonomic Description: Large climbers, with hollow, striate branches. Leaves orbicular-cordate, lobed, serrate. Flowers reddish-brown, in pyramidal thyrsoid cymes, berries globose, black, 2-seeded.

Medicinal Parts: Root.

Chemical Constituents: Root contains Saponins

Medicinal Properties: Root - Astringent.

Medicinal Uses: Root - Chronic dysentery, Wounds.

Leaf - Dental problem, Indolent ulcers, Wounds.

Ethno- Medicinal Uses: Root - Boils (Tarafder, 1984); bone fracture (Joshi *et al.*, 1980; Saxena, 1986; Shah & Gopal, 1986); dysentery (Malhotra & Moorthy, 1973; Sinha *et al.*, 2006); inflammation (Tribedi *et al.*, 1985; Sinha *et al.*, 2006); stomach pain (Joshi *et al.*, 1980); Skin diseases (Joshi *et* *al.*, 1980; Shah & Gopal, 1986; Singh *et al.*, 1994; Saxena, 1986; Sinha *et al.*, 2006); snake bite (Mishra & Billore, 1983; Murthy *et al.*, 1986); Corns (Sharma *et al.*, 1985); Wounds (Shah *et al.*, 1983; Saxena & Dutta, 1975; Chandra *et al.*, 1989); Menstrual trouble (Shah *et al.*, 1981); Rheumatism (Jain & Puri, 1984); Gonorrhoea, Syphilis (Sharma, 1988); Diarrhoea (Sinha *et al.*, 2006).

Stem - Bone fracture (Singh *et al.*, 1994).

Stem bark - Stomach pain (Joshi *et al.*, 1980).

ANGELICA GLAUCA Edgew. (APIACEAE)

Regional & Other Names : *English* – Long wort; *Hindi* – Churai; *Sanskrit* – Coraka.

Distribution: Western Himalaya, from Kashmir to Shimla 8,000-10,000 ft.

Taxonomic Description: Shrub, paripinnate compound leaf; Violet flowers in umbel inflorescence; cremocarp fruit.

Medicinal Parts: Roots.

Chemical Constituents: Roots contains Essential oil.

Medicinal Properties:

चोरको मधुरस्तिक्तः कटुः पाके कटुर्लघुः ।
तीक्ष्णो हृघोऽहिमो हन्ति कुष्ठकण्डूकफानिलान् ।।
रक्षोऽश्रीस्वेदमेदोऽस्त्रज्वरगन्धविषव्रणान् ।। (भा. प्र.)

Cōrakō madhurstiktaḥ katuḥ pāke katurlaghuḥ
Tīksnō hrdhoahimō hanti kuṣthakandukaphānilān
Rakshoashrisvedmedōastrjvaragandhavisavrināṇ.(Bha. Pra.)

Root - Aromatic, Cardiotonic, Carminative, Diaphoretic, Diuretic, Expectorant, Pungent, Purgative, Stimulant, Stomachic, Sudorific, Tonic.

Medicinal Uses: Root - Anorexia, Biliousness, Bronchitis, Dyspepsia, Flatulence, Stomachache, Typhoid, Wounds.

Ethno- Medicinal Uses: Root - Stomach complaint, Stops vomiting (Shah & Joshi, 1981); dyspepsia, stomachache, vomiting (Singh & Aswal,

1992; Singh, 1993; Dobriyal *et al.,* 1997); constipation (Singh & Aswal, 1992; Singh, 1993).

Px - Dysentery; Menorrhoea (Jain & Tarafder, 1970).

Rhizome - Gastric complaint (Rawat & Pangtey, 1987).

Leaves- Stimulant, cordial & stomach troubles (Maikhuri *et al.,* 1998).

ANISOMELES INDICA (L.) Kuntze. (LAMIACEAE)

Syn.: *A. ovata* R. Br.

Regional & Other Names : *Gujarati* -Gopali; *Kannada* –Mangamari soppu; *Malayalam* -Chedayan; *Tamil* –Chedayan.

Distribution: Throughout India.

Taxonomic Description: Suffruticose herbs, with ovate, acute or obtuse, crenate-serrate, softe, softly pubescent leaves. Flowers bluish-purple. Nutlets broadly ovoid, with rounded ends, polished.

Medicinal Parts: Whole plant.

Chemical Constituents: The plant contains Essential oil.

Medicinal Properties- The plant has astringent, carminative, tonic properties.

Medicinal Uses: Essential oil from plant - Uterine affections.

Ethno-Medicinal Uses: Whole plant - Astringent, Carminative (Sharma *et al.,* 1979); Refrigerant (Bhatt & Sabnis, 1987); Antipyretic (Bhatt & Sabnis, 1987); fever, nervous complaint (Jain & De, 1966); muscular pain (Pal, 1984); To make hair silky (Gopal & Chandra, 2003); To prevent falling of hairs (Ghosh, 2006).

Aerial parts - Burning sensation during urination (Purkayastha & Nath, 2006).

ARTEMISIA MARITIMA L. (ASTERACEAE)

Syn.: *A. fragrans* (Willd.) Ledeb.; *A. spicigera* C. Koch.; *A. brevifolia* Wall.

Regional & Other Names : *English* – Worm seed; *Gujarati* - Chhuvari-ajamoda; *Hindi* – Ajavayana; *Marathi* –Kirmaniova; *Sanskrit* – Yavani.

Distribution: Western Himalaya, from Kashmir to Kumoan 7,000-11,000 ft.

Taxonomic Description: A deciduous shrubs, with rough, fibrous Bark and Yellowish Homogenous heads.

Medicinal Parts: All parts.

Chemical Constituents: Essential oil, Santonin, β-Santonin, Pseudosantonin and a bitter Substance Artemisin.

Medicinal Properties:

यवानिका यवानी स्याच्चौहारो जन्तुनाशनः ।
चौराहस्तद्गुणः प्रोक्तो विशेषात् कृमिनाशनः ।। (रा. नि.)

Yavānikā yavāni ̣syāçcaūhārō jaṇtunāsanāḥ
Caūrāhaṣtaḍguṇaḥ p̣rokto visesāṭ kṛminãsanaḥ. (Rā. Ni.)

Seed - Anthelmintic, Aphrodisiac, Appetiser, Bitter, Hot, Laxative, Pungent, Stomachic, Tonic.

Medicinal Uses: Seed - Abdominal pain, Fever, Indigestion, Mucous diarrhoea.

Ethno- Medicinal Uses: Whole plant - Anthelmintic, Astringent, Cuts, Gastric complaint (Gupta *et al.,* 1981); Blood purifier (Baruah *et al.,* 1987).

Leaf - Stomach problems (Ballabh & Chaurasia, 2006); Intestinal worms (Chaurasia *et al.,* 2003).

ARTOCARPUS COMMUNIS Forst. (MORACEAE)

Syn.: *A. incisus* (Thunb.) L. f.; *A. altilis* (Park.) Fosberg.

Regional & Other Names : *English* – Bread fruit; *Hindi* – Bilatti Kathal; *Malayalam* - Katappilavu; *Marathi* – Vilayati phanas; *Sanskrit* – Ksudrapanasah; *Tamil* - Simaippala; *Telugu* – Simapavara.

Distribution: Cultivated throughout India.

Taxonomic Description: Tall evergreen trees, with catkin inflorescence and prickly, Globose composite fruits.

Medicinal Parts: Fruits, Latex.

Chemical Constituents: Fruits contain Iodine, Fluorine.

Medicinal Properties:

> मधुरं कटुकं क्षुद्रपनसस्य फलं स्मृतम् ।
> मूत्रलं स्तन्यजननं पित्तास्रज्वरदाहनुत् ।।
>
> Madhuraṃ katukaṃ ksudrapanaṣasya phalaṃ ṣmṛtaṃ
> Mūṭralaṃ ṣtaṇyajananaṃ piṭtāsrajvaradāhanuṭ.

Fruits - Acrid, Appetiser, Bitter, Cardiotonic, Cooling, Diuretic, Stomachic, Sweet.

Medicinal Uses: Fruits - Cardiac debility, Colic, Flatulence.

ARTOCARPUS HIRSUTUS Lam. (MORACEAE)

Regional & Other Names : *Bengali* – Katelgach; *English* – Wild jack; *Gujarati* - Panas; *Hindi* – Katahal; *Kannada* - Hebpalasu; *Malayalam* - Ayani; *Marathi* – Phanas; *Sanskrit* – Lakucah; *Tamil* - Kattuppala; *Telugu* – Pejuta.

Distribution: Evergreen forests of West Coast.

Taxonomic Description: Evergreen tall trees, with ovate leaves and spinous, cylindrical syconous fruits.

Medicinal Parts: Bark, Leaves, Fruits.

Chemical Constituents: Fruit contains β-Carotene, Cycloartenone, Cycloartenol, β-Sitosterol, Butyrospermol, Leusine, Isoleucine, Tyrosine, Valine.

Bark contains Betulinic acid, Cycloheterophyllin, Isocycloheterophyllin, Cycloartenyl acetate, Cycloartenol, Cycloartenone.

Leaves contain Cycloartenone, Cycloartenol, β -Sitosterol, Tannins.

Medicinal Properties:

लचुकः स्वरसे तिक्तः कषायोष्णो लघुस्तथा ।
कफदोषहरो दाही मलसंग्रहदायकः ।। (रा.नि.)

Lacukaḥ ṣvarase tiktaḥ kasāyoṣṇo laghuṣtathā
Kaphadosaharo dāhī malasaṃgrahadāyakaḥ. (Ra. Ni.)

Ripe fruits - Aphrodisiac, Appetiser, Cooling, Indigestible, Sour, Sweet.

Raw fruits - Aphrodisiac, Astringent, constipating, Indigestible, Sour, Sweet, thermogenic.

Medicinal Uses: Leaves – Hydrocele.

Raw fruits- Colic.

Ripe fruits - Anorexia.

Bark- Small pimples, Cracks on the Skin (External application).

Root - Bloody dysentery.

Ethno-Medicinal Uses: Seed -Rheumatism, for hair growth (John, 1984).

ARUNDO DONAX L. (POACEAE)

Regional & Other Names : *Bengali* – Gaha nal; *English* – Spanish cane; *Hindi* – Bara nal; *Malayalam* - Ama; *Sanskrit* – Nala; *Tamil* – Korukkai; *Telugu* – Advani-Keakasagaddi.

Distribution: The plant is found wild and cultivated in lower Himalaya to Kashmir, East wards to North-Eastern India upto 2400 m. altitude, extending to South India, except West Coast.

Taxonomic Description: Perennial plant, with light greenish yellow flowers and caryopsis fruits.

Medicinal Parts: Rhizome.

Chemical Constituents: Rhizome contains - Indole bases, Dehydro-bufotenine, Bufotenidine, Gramine (Donaxine).

Medicinal Properties:

नलस्तु मधुरस्तिक्तः कषायः कफरक्तजित्।। (भा. प्र.)

Nalastu madhurstiktaḥ kasāyaḥ kapharaktajit. (Bha. Pra.)

Rhizome - Aphrodisiac, Cooling, Diuretic, Emollient, Galactagogue, Hypotensive, Sudorific, Sweet.

Medicinal Uses: Rhizome - Blood diseases, Dysuria, Erysipelas, Eye diseases, Leucorrhoea, Piles.

Ethno-Medicinal Uses: Plant- Urinary disorders, ulcers, tonic (Srivastava *et al.,* 1986).

ATALANTIA MONOPHYLLA Corr. (RUTACEAE)

Syn.: *A. malabarica* (Rafin.) Tanaka.

Regional & Other Names : *English* – Wild lime; *Hindi* – Jungli nimbu; *Kannada* – Kadu-nimbo; *Malayalam* - Malanarakam; *Marathi* – Makad limbu; *Sanskrit* – Atavijambira; *Tamil* – Perungu-ngrungan; *Telugu* – Adavi nimma rundu.

Distribution: Konkan, S.M. Country, North Kanara, Madras Presidency, Western Coast, Western Ghats, Deccan, North Circars, Ceylon, Burma, Malay Peninsula. Indo-China, Java.

Taxonomic Description: A shrub or treelet about 6 m. tall, with short sharp spines, leaves lanceolate to ovate; Flowers white in Raceme or Cyme Inflorescence; fruits green, Globose Berries.

Medicinal Parts: Berries, Leaves, Root.

Chemical Constituents: Root bark contains a Tetranortriterpenoid, Atalantin; Stigmasterol, Xanthyletin, Marmesin, Y-Sitosterol; Acridone Alkaloid.

Fruits contain Alkaloid serverine.

Medicinal Properties:

जम्बीरस्यफलं रसेम्लमधुरं वातापहं पित्तकृत् ।
पथ्यं पाचनरोचनं वलकरं वन्हेर्विवृद्धिप्रदम् ।। (रा.नि.)

Jaṃbīraṣya phalaṃ rasemlamadhuraṃ vātāpahaṃ piṭtakṛt
Paṭhyaṃ pācanarocanaṃ balakaraṃ vaṇheṛvivṛddhipṛadaṃ.
(Rā. Ni.)

Root - Antiseptic, Antispasmodic, Stimulant.

Medicinal Uses: Root- Rheumatism, Swellings.

Ethno- Medicinal Uses: Leaf - Dysentery (Upadhye *et al.,* 1986).

BRASSICA NIGRA (L.) Koch. (BRASSICACEAE)

Syn.: *Sinapis nigra* L.; *S. erysimoides* Roxb.

Regional & Other Names : *Bengali* – Raisarisha; *English* – Black mustard; *Gujarati* - Rai; *Hindi* – Rai; *Kannada* – Kari soasivae; *Malayalam* - Katuka; *Sanskrit* – Asuri; *Tamil* - Kadugu; *Telugu* – Avalu.

Distribution: Cultivated throughout India.

Taxonomic Description: A rigid, Branching annual upper leaves entire. Flowers bright yellow, in nacked racemes. Fruits siliqua.

Medicinal Parts: Leaves, Seeds, Seed oil.

Chemical Constituents: Seed contains Glucoside Singirin, Sinapine & a Volatile Isothiocyanate and Essential oil.

Medicinal Properties:

सर्षपस्तु रसे पाके कटुः स्निग्धः सातिक्तकः ।
तीक्ष्णोष्णः कफवातघ्नो रक्तपित्ताभिवर्धनः ।। (भा.प्र.)

Saṛsapaṣtu rase pāke katuh ̣snigdhaḥ sātiktakaḥ
Tīkṣṇoṣṇaḥ kaphavātaghno raktapiṭtābhivaṛdhanaḥ. (Bhā. Pṛa.)

Seeds- Appetiser, Hot Sharp bitter taste.

Medicinal Uses: Seeds - Anthelmintic, Appetiser, Cough, Enlargement of spleen, Itching, Skin diseases, Tumours.

Ethno- Medicinal Uses: Seed oil- Rheumatism (Banerjee, 1977).

Seed - Bronchitis, Fever (Rajan, 1992); Hay fever, Scurvy, Coryza, Cough, Diarrhoea (Rajan *et al.,* 2003).

CAESALPINIA DIGYNA Rottl. (CAESALPINIACEAE)

Syn.: *C. oleosperma* Roxb.

Regional & Other Names : *Bengali* – Umulkuchi; *English* – Teri pods; *Hindi* – Vakerimul; *Marathi* – Vakerimula; *Sanskrit* – Vakeri; *Telugu* – Nunegacha.

Distribution: In Konkan, Bengal, Assam & Andamaan.

Taxonomic Description: A large scandent prickly perennial climbing herb.

Medicinal Parts: Leaves, Roots & Nodes on Root.

Chemical Constituents: Tannin in Pod-cases but not in seeds; Bitter substance bonducin, Saponin.

Medicinal Properties:

वाकेरी तु त्रिदोषघ्नी शोषघ्नी तुवरा परम् ।
शस्यते गण्डमालायां प्रमेहे राजयक्ष्मणि ।। (द्र. वि.)

Vākerī tu tridosaghnī ṣosaghnī tuwarā paraṃ
Śaṣyate gandamālāyāṃ pṛamehe rājayaksmaṇi (Dṛa. Vi.)

Root - Astringent.

Medicinal Uses: Root – Diabetes, Diarrhoea, Goitre.

CAJANUS CAJAN (L.) Millsp. (FABACEAE)

Syn.: *C. indicus* Spreng.

Regional & Other Names : *Bengali* – Adahar; *English* – Pigeon pea; *Gujarati* - Tuvar; *Hindi* – Tuvar; *Kannada* - Togari; *Malayalam* - Tuvara; *Marathi* – Tuvar; *Sanskrit* – Adhaki; *Tamil* - Tuvarai; *Telugu* – Kandipappu.

Distribution: Cultivated throughout India.

Taxonomic Description: An erect shrub. Leaflets 3, oblong-lanceolate. Flowers yellow, in corymbose racemes. Pods narrowed at the ends, torulose.

Medicinal Parts: Leaves, Seeds.

Chemical Constituents: Leaves contain n-Alkanes, Simple esters, Free alcohols, Lupeol, Phytosterols as β-Sitostrerol, Stigmasterol, Campesterol, Cholesterol; Isoflavone as Cajanol; Salicylic acid, Linoleic acid, Oleic acid.

Medicinal Properties:

आढकी तुवरा रूक्षा मधुरा शीतला लघुः ।
ग्राहणी वातजननी वर्ण्या पित्तकफास्रजित् ।। (भा. प्र.)

Ādhaki tuvarā rukṣā madhurā śītalā laghuḥ
Gṛāhiṇi vātajanani vaṛṇyā piṭtakaphāsṛajiṭ. (Bhā. Pṛa.)

Leaves - Anti-inflammatory, Astringent, Anodyne, Cooling, Diuretic, Laxative, Sweet.

Seeds - Acrid, Alexeteric, Anthelmintic, Astringent, Cooling, Constipating, Febrifuge, Resolvent, Sweet.

Medicinal Uses: Leaves / Seeds - Bronchitis, Cardiac diseases, Cough, Fever, Haemorrhoids, Intestinal Worms, Tumours, Vomiting.

Ethno-Medicinal Uses: Leaves - Gargle (Shah *et al.,* 1981); Skin diseases (Singh *et al.,* 1987); Cholera (Shukla *et al.,* 1992); Burnt skin (Singh, 2003); Eye irritation (Bora, 2003); Jaundice (Baruah *et al.,* 2006;Das *et al.,* 2006); Diabetes (Ghosh, 2006).

Seeds - Gout (Jadeja *et al.,* 2006); Diabetes (Ghosh, 2006).

Root- Diabetes (Ghosh, 2006).

CALAMUS ROTANG L. (ARECACEAE)

Regional & Other Names : *Bengali* – Bet; *English* – Common rattan; *Gujarati* - Netar; *Hindi* – Bemt; *Kannada* - Betasu; *Malayalam* - Cural; *Marathi* – Bet; *Sanskrit* – Vetrah; *Tamil* - Pirampu; *Telugu* – Jatayurkuli.

Distribution: Central and South Indian states, in dry tracts upto 450 m. altitude.

Taxonomic Description: Scandent slender shrubs, with unisexual flowers and subglobose fruits.

Medicinal Parts: Whole plant.

Chemical Constituents: Yellow or Greenish yellow volatile oil containing cineol, Pinene, Borneol, Ketones & Esters.

Medicinal Properties:

वेत्रस्तु तुवरः शीतः तिक्तः कटु कफापहः ।
वातं पित्तं च दाहञ् च शोफार्शोऽश्मरिकृच्छ्रकान् ।। (नि. र.)

Vetṛaṣtu tuvaraḥ śītaḥ tiktaḥ katu kaphāpahaḥ
Vātaṃ pịttaṃ ca dāhaṃ ca śophāṛśoaṣmarikṛcchrakāṇ. (Ni. Ra.)

Root- Acrid, Astringent, Alexeteric, Anti-dysenteric, Anti-inflammatory, Bitter, Cooling, Expectorant, Depurative, Diuretic, Febrifuge, Tonic.

Leaves- Acrid, Bitter, Cooling, Astringent, Laxative.

Seed -Acrid, Astringent Depurative, Sour, Expectorant.

Medicinal Uses: Root - Bronchitis, Burning sensation, Cough, Dysentery, Erysipelas, General debility, Inflammation, Leprosy, Skin diseases, Strangury.

Leaves - Leprosy, Pruritus, Skin disease.

Seed - Cough, Pruritus, Skin disease.

Ethno- Medicinal Uses: Px- Convulsions (Jain & Tarafder, 1970).

CANAVALIA GLADIATA (Jacq.) DC. (FABACEAE)

Syn.: *C. ensiformis* Sensu Baker non DC.

Regional & Other Names : *Bengali* – Makhan shim; *English* – Borad bean; *Hindi* – Lalkudusumpal; *Kannada* - Sembi; *Malayalam* - Valavara; *Marathi* – Abadi; *Sanskrit* – Asisimbi; *Tamil* - Segapputampattai; *Telugu* – Tamma.

Distribution: Cultivated throughout India.

Taxonomic Description: A perennial twiner, with white or lilia flowers and sword shaped flattened pods.

Medicinal Parts: Fruits.

Chemical Constituents: *Seeds contain* Crude Protein 29%, Crude Fiber 6%, Starch 37%, Ether extract 1.5%, Nitrogen free extract 62% and total Sugar 7.5%. The amino acid contents in the protein fraction were: Lycine 6.5%, Histidine 4%, Arginine 6%, Tryptophan 2%, Treoline 6%, Cistine 0.8%,

Valine 6%, Methionine 0.5%, Isoleucine 5%, Leucine 9%, Tyrosine 2% and Phenylalanine 5%.

Medicinal Properties:

आसिशिम्बी तु मधुरा कषाया श्लेष्मपित्तजित् ।
व्रर्णदोषापहन्त्री च शीतला रूचिदीपनी ।। (रा. नि.)

Asiṣiṃbī tu madhurā kaṣāyā śleṣmapịttajiṭ
Vṛanadosāpahaṇṭṛī ca śītalā rucidīpanī. (Rā. Ni.)

Fruits - Appetiser, Astringent, Cooling, Digestive, Sweet, Vulnerary.

Medicinal Uses: Fruits - Anorexia, Burning Sensation, Dyspepsia, Ulcers, Wounds.

Ethno- Medicinal Uses: Root - Intestinal worms (Sudhakar & Rolla, 1985)

Leaf - Piles (Ganesan *et al.,* 2006).

CANSCORA DECUSSATA (Roxb.) Schult. (GENTIANACEAE)

Regional & Other Names : *Bengali* – Dankuni; *English* – Canscora; *Hindi* – Samkhaphuli; *Kannada* - Sankinisopppu; *Malayalam* - Samkha-puspi; *Marathi* – Sankhapushpi; *Sanskrit* – Sankhini; *Telugu* – Kancakura.

Distribution: Throughout India.

Taxonomic Description: Dichotomously branched annual, with quadrangular Stem and White flowers.

Medicinal Parts: Whole plant.

Chemical Constituents: Loliolide, Triterpenes as Glucose Conscoradione, Friedelin, Canscoridine, Sterols as-Sitosterol, Stigmasterol, Campesterol.

Medicinal Properties:

शङ्.खपुष्पी सरा मेध्या वृष्या मानसरागहृत् ।
रसायनी कषायोष्णा स्मृतिकान्तिबलाग्निदा ।
दोषापस्मारभूताश्रीकुष्ठकृमिविषप्रणुत् ।। (भा. प्र.)

Śaṇkhapuṣpī sarā medḥyā vṛṣyā mānasarogahṛt
Rasāyanī kasāyoṣṇā ṣmṛtikāṇtibalāgnidā
Dosāpaṣmārabhūtāsrīkuṣthakṛmivisapṛaṇuṭ. (Bhā. Pṛa.)

Plant - Acrid, Anthelmintic, Aphrodisiac, Appetiser, Astringent, Bitter, Emollient, Elexeteric, Depurative, Laxative, Thermogenic, Tonic, Vulnerary.

Medicinal Uses: Plant - Abdominal disorders, Ascites, Epilepsy, General debility, Inflammation, Intestinal worms, Leprosy, Leucoderma, Nervous debility Skin diseases, Tuberculosis, Ulcers.

Ethno- Medicinal Uses: Stem - Blood purifier, Stomachache (Srivastava & Varma, 1981).

Whole plant - Flatulence (Singh *et al.,* 1992); Stomachache, Blood purifier (Sinha *et al.,* 2006); Nervous debility & Epilepsy (Basak, 2006).

CANTHIUM PARVIFLORUM Lam. (RUBIACEAE)

Regional & Other Names : *English* – Carray cheddie; *Hindi* – Kirma; *Kannada* - Kakegida; *Malayalam* - Cerukara; *Marathi* – Kirma; *Sanskrit* – Gangeruki; *Tamil* - Karaicceti; *Telugu* – Balusa.

Distribution: Throughout India.

Taxonomic Description: A thorny shrub, with Yellow flowers and Drupe fruits.

Medicinal Parts: Root, Leaves.

Chemical Constituents: Mannitol & Alkaloids.

Medicinal Properties:

गाङ्.गेरूकी कषायोष्णा मधुरा ग्राहि मूत्रला ।।

Gāṇgerūkī kasāyoṣṇā madhurā gṛāhi mūṭralā.

Root / Leaves - Anthelmintic, Astringent, Diuretic, Febrifuge, Sweet, Constipating, Thermogenic, Tonic.

Medicinal Uses: Root / Leaves - Diarrhoea, Fever, General debility, Intestinal worms, Leucorrhoea, Strangury.

Ethno- Medicinal Uses: Fruits - Headache (Saxena *et al.,* 1981).

Root - Swell neck (Saxena *et al.,* 1981).

Leaf - Fever (Prasad *et al.,* 2003); Swelling, Inflammation, Dysentery (Rosakutty *et al.,* 2003).

CAPPARIS ZEYLANICA L. (CAPPARACEAE)

Syn.: C. horrida L. f.

Regional & Other Names : *Bengali* – Kalokera; *English* – Ceylon caper; *Hindi* – Ardanda; *Kannada* - Anatundikayee; *Marathi* – Govindi; *Sanskrit* – Karambha; *Tamil* - Kattotti; *Telugu* – Doddi.

Distribution: Throughout India.

Taxonomic Description: Climbing shrubs; leaves ovate or oblong, entire; stipular spines hooked; petals densely hairy inside; berries subglobose, reddish-brown, with woody pericarp; seeds subglobose.

Medicinal Parts: Root Bark, Leaves, Fruits.

Chemical Constituents: An alkaloid- □-Phytosterol, □-Mucilaginous Substance and water soluble acid. L-Stachydrine, Rutin and β -Sitosterol.

Seeds / Leaves - Thioglucoside, Glucocapparin, n-Tricontane, □-amyrine & Fixed oil.

Medicinal Properties:

गान्धारी कटुतिक्तोष्णा कफवातनिकृन्तनी ।
शोफघ्नी दीपनी रूच्या रक्तग्रंथिरूजापहा ।। (रा. नि.)

Gāṇdhārī katutiktoṣṇā kaphavātānikṛṇtanī
Śophaghnī dīpanī rūcyā raktagṛaṇthirujāpahā. (Rā. Ni.)

Root Bark – Appetiser, Demulcent, Sedative, Stomachic.

Medicinal Uses: Root Bark - Colic.

Ethno- Medicinal Uses: Root Bark - Anthelmintic (Sharma *et al.,* 1979); Boils (Singh, 2003; Maheshwari *et al.,* 1980; Maheshwari *et al.,* 1981); Cholera (Sharma *et al.,* 1979); Toothache (Shivakumar *et al.,* 2003); Swelling of testicles (Singh, 2003; Prakash & Singh, 2006); Cholera, Piles, Anthlemintic (Sinha *et al.,* 2006); Cough & breast pain due to excessive lactation in women (Kothari & Rao, 2003).

Leaf - Piles (Sinha *et al.,* 2006; Sebastian & Bhandari, 1984; Sharma *et al.*, 1979); Swellings (Sharma *et al.,* 1979); Cholera, Anthlemintic (Sinha *et al.,* 2006).

Fruits - Worms in stomach (Ganesan & Kesavan, 2003); Rheumatism, Snake bite (Panda & Das, 2003); Snake bite (Singh *et al.,* 1987).

Young stem - Swellings, Cough, Asthma (Deokule, 2006).

Root - Cholera, Piles, Anthlemintic (Sinha *et al.,* 2006); Dyspepsia (Shivakumar *et al.,* 2003).

CARISSA CARANDAS L. (APOCYNACEAE)

Syn.: *C. cognesta* Wt.

Regional & Other Names : *Bengali* – Karamcha; *English* – Christ's thorn; *Gujarati* -Karundo; *Hindi* -Karaunda; *Kannada* –Karekai; *Malayalam* -Kala; *Marathi* –Karavanda; *Sanskrit* - Avighnah; *Tamil* - Karanta; *Telugu* – Kalivi.

Distribution: Throughout India.

Taxonomic Description: Erect or scandent shrubs, with paired spines at each node. Leaves elliptic-oblong to ovate. Flowers white, in corymbose cymes. Berries elliptic-oblong, fleshy.

Medicinal Parts: Root, Fruits.

Chemical Constituents: Root contain - β -sitosterol, a Sesquiterpene as Carissone: a Lignancarinol, Odoroside-H, Digitoxigenin, Glucose, D-digitalose, Lupeol, Ursolic acid, Methyl ursolate.

Fruits contain - Triterpene alcohol as Carissol, Glucose, Galactose, Serine, Glutamine, Alanine, Valine, Glycine, Carissic acid, Phenylalanine.

Medicinal Properties:

करमर्दः सतिक्ताम्लो बालो दीपनदाहकः ।
पक्वस्त्रिदोषशमनोऽरुचिघ्नो विषनाशनः ।। (रा. नि.)

Karamaṛdaḥ satiktāṃlo bālo dīpanadāhakaḥ
Pakvaṣṭridosaṣamanoarucighno visanāsanaḥ. (Rā. Ni.)

Leaves - Febrifuge.

Roots - Anthlemintic, Antiscorbutic, Stomachic.

Ripe fruits – Antiscorbutic, Appetiser, Cooling, Sweet.

Unripe fruits - Antipyretic, Aphrodisiac, Appetiser, Astringent, Bitter, Constipating, Sour, Thermogenic.

Medicinal Uses: Roots - Intestinal worms, Pruritus, Scabies, Stomach disorders.

Ripe fruits - Burning sensation, Skin diseases, Scabies, Pruritus.

Unripe fruits - Anorexia, Diarrhoea, Intermittent fever.

Ethno- Medicinal Uses: Leaf - Whooping cough (Verma & Chauhan, 2006); Asthma (Kothari & Rao, 2003); Fever (Nautiyal, 1981; Verma & Chauhan, 2006).

Seed - Itching (Gupta, 1981).

Fruits - Scurvy (Dwivedi, 2003); To promote appetite (Shukla *et al.*, 1992); Cough (Verma & Chauhan, 2006).

Root bark - Cancer (Ganesan *et al.*, 2006); Cough (Verma & Chauhan, 2006); Wounds (Nautiyal, 1981).

Root- Skin diseases (Kothari & Rao, 2003); Wounds (Nautiyal, 1981).

CASSIA ALATA L. (CAESALPINNIACEAE)

Regional & Other Names : *Bengali* – Dadamari; *English* – Ringworm Senna; *Hindi* – Dadmurdan; *Malayalam* - Elakajam; *Marathi* – Dadamardana; *Sanskrit* – Dadrughna; *Tamil* - Anjlai; *Telugu* – Mettatamara.

Distribution: A native of West Indies. Introduced into India.

Taxonomic Description: A large shrubs, with Yellow flowers in pedunculate racemes inflorescence and long, ligulate with a broad wing down the middle each valve, membranous, dehiscent, straight & glabrous pods.

Medicinal Parts: Leaves, Flowers.

Chemical Constituents: Plant contains Chrysophanic acid; Chrysophenol, Emodin, Rhein and Aloe-emodi.

Leaves contain Kaempferol and Aloe-emodin and a Volatile oil (Sesquiterpene and Phenolic Compounds).

Roots contain Quinone Pigments.

Seeds yield Galactomannan; Emodin, Aloe-emodin, β-Sitosterol.

***Medicinal Properties*:**

चक्रमदर्दो लघुः स्वादु रूक्षः पित्तानिलापहः ।
हृद्यो हिमः कफश्वास कुष्ठदद्रुक्रिमीन हरेत् ।। (द्रव्यगुण)

Cakramadaṛdo laghuḥ ṣvādu ruksaḥ piṭtānilāpahaḥ
Hṛdyo himaḥ kaphaṣvāsakuṣthadadṛukṛīmin hareṭ. (Dṛavyaguṇa)

Leaves - Antifungal, Antiparasitic, Astringent, Purgative, Sour.

Medicinal Uses: Leaves – Asthma, Bronchitis, Eczema, Ringworm, Skin diseases, Snake bite.

Flowers – Atshma, Bronchitis.

Ethno- Medicinal Uses: Leaf - Skin diseases (Rao & Jamir, 1982; Tiwari *et al.,* 1979; Sharma & Baishya, 1992; Das *et al.,* 2006); Ringworm (Sharma & Baishya, 1992; Girach *et al.,* 2003; Das *et al.,* 2006); Eczema (Tiwari *et al.,* 1979; Rao & Jamir, 1982; Purkayastha & Nath, 2006; Sinha *et al.,* 2006); Itching (Girach *et al.,* 2003); Allergy (Das *et al.,* 2006) .

Flower & Leaves - Ringworm (Sharma, 2003).

CASSIA SENNA L. (CAESALPINNIACEAE)

Syn.: *C. angustifolia* Vahl.; *Senna officinalis* Roxb.

Regional & Other Names : *Bengali* – Sanna-makki; *English* – Indian senna; *Gujarati* – Nat-ki-sana; *Hindi* – Sanay; *Kannada* - Sonamukhi; *Malayalam* - Sunnamukki; *Marathi* – Shona-makhi; *Sanskrit* – Svarnapatri; *Tamil* - Nilavirai; *Telugu* – Nela tangedu.

Distribution: Cultivated in Tamil Nadu.

Taxonomic Description: A small shrub. Leaves paripinnate; Leaflets glabrous, yellowish Green. Flowers yellow. Pods greenish-brown to dark Brown.

Medicinal Parts: Leaves.

Chemical Constituents: Fruits contain Sennosides A & B, Sennidin-8, 8-Diglucoside, Oxymethyl-antrhaquinone.

Leaves contain Flavonol, Isorhamnetin, Kaempferol, Rehin, Emodine, Sennosides A,B,C,D.

Medicinal Properties:

रूक्षोष्णा शोधनी बाढं वातश्लेष्मविरोधिनी ।
शस्ता सनायकी नाम मनाक् पीततनुच्छदा ।। (सि.भे.म.)

Rūksoṣṇā śodhanī bādhaṃ vātaśleṣmavirōdhinī
Śaṣtā sanāyakī nāma manāk pītatanucchadā. (Si. Bhe.Ma.)

Leaves - Acrid, Astringent, Anthelmintic, Bitter, Depurative, Febrifuge, Expectorant, Sweet, Thermogenic.

Medicinal Uses: Leaves - Abdominal disorders, Bronchitis, Cough, Anaemia, dyspepsia, Jaundice, Helminthiasis, Leucoderma, Leprosy, Skin diseases, Constipation.

Ethno- Medicinal Uses: Root - Joint pain (Rajendra & Mehrotra, 1996).

Leaves - Habitual constipation (Singh & Pandey, 1998).

CERBERA ODOLLAM Gaertn. (APOCYNACEAE)

Syn.: *C. manghas* L.

Regional & Other Names : *Bengali* – Dabur; *English* – Odallum tree; *Hindi* – Dabur; *Kannada* – Monde; *Malayalam* - Otalam; *Marathi* – Sukanu; *Sanskrit* – Auddalakah; *Tamil* – Udalai; *Telugu* – Adavi arali.

Distribution: Throughout India.

Taxonomic Description: An evergreen tree, with Milky juice; Leaves lanceolate or oblanceolate, glabrous; flowers large, White with a Yellow throat; fruits drupes with fibrous pericarp.

Medicinal Parts: Bark, Leaves, Fruits, Milky juice.

Chemical Constituents: A poisonous glycoside identical with thevatine; cerbenin occurs in the seeds which yield 55% of a fixed oil and ash 3.3% bitter substances odollin.

Medicinal Properties:

औद्दालकस्य फलं तीक्ष्णं अलर्क विषनाशकम् ।। (स्व.)

"Aūḍdālakaṣya phalaṃ tīksnaṃ alaṛkviśanāṣakaṃ." (Ṣva.)

Bark/Leaves/ Milky Juice - Purgative, Emetic.

Fruits- Poisonous, Narcotic.

Medicinal Uses: Bark - Ringworm infestation.

Leaves - Skin diseases in children.

Fruits- Hydrophobia.

CHENOPODIUM ANTHELMINTICUM Bert. (CHENOPODIACEAE)

Syn.: *C. ambrosioides* L.

Regional & Other Names : *English* – Worm seed; *Hindi* – Khatu; *Kannada* – Kadu voma; *Malayalam* - Katu Ayamodakam; *Sanskrit* – Svetachilli; *Tamil* - Kattasambadan.

Distribution: Throughout India.

Taxonomic Description: An erect, strongly foetid-smelling, wolly-puberculous, much-branched shrub, often gregrarious, pale green in appearance, stems ribbed, often purple-tinged, leafy. Leaves oblong-ovate, sinuate-dentate. Flowers greenish, some turning to purplish in small clusters forming slender, axillary, interrupted and terminal simple or paniculate spikes. Perianth 5-lobed, clasping the fruit. Seed brown polished, smooth.

Medicinal Parts: Root, Leaves, Flowers, Seeds.

Chemical Constituents: The plant contains an alkaloid Chenopodine; Amino acids as alanine, Valine, Leucine, Glycine; Carboxylic acids as Oxalic, Malic & Succinic, acid.

Leaves contain Flavonoids as Kaempferol-7-rhamnoside & Ambroside.

Seed contains A Volatile oil & Triterpenoid saponins, A Scaridole, Cineol, P-cymene, □-Prpinene, Dihydro-P-cymene, Betazine, Choline & Sylvestrene.

Medicinal Properties:

श्वेतचिल्ली सुमधुरा क्षारा च शिशिरा च सा ।
त्रिदोषशमनी पथ्या ज्वरदोषविनाशिनी ।। (रा. नि.)

Ṣvetachiḷḷī sumadhurā ksārā ca śiśirā ca sā
Tṛidosa śamanī paṭhyā jvardosavināśinī. (Rā. Ni.)

Root / Leaves - Abortifacient, Analgesic, Antiasthmatic, Antirheumatic, Antispasmodic, Antitumour, Carminative, Diaphoretic, Diuretic, Emenagogue, Febrifuge, Stomachic.

Seed / Seed oil - Anthelmintic, Insecticidal.

Medicinal Uses: Leaves / Roots - Amenorrhoea, Anaemia, Asthma, Bronchitis, Colic, Cough, Dysentery, Fever, Flatulence, Intestinal disorders, Nervous disorders, Rheumatism, Stomach pain, Urinary disorders.

Seed / Seed oil - Cholera, Hysteria, Intestinal worms, Malaria, Nervous diseases.

Ethno- Medicinal Uses: Leaf - Anthelmintic (Singh, 1945); Influenza, Pneumonia, Typhoid (Rao, 1981).

CHONEMORPHA FRAGRANS (Moon.) Alston. (APOCYNACEAE)

Syn.: *C. macrophylla* (Roxb.) G. Don.

Regional & Other Names : *Hindi* – Garbhedaro; *Kannada* – Manjinaru; *Malayalam* - Perunkurumpa; *Sanskrit* – Murva; *Tamil* – Parumkurumpa; *Telugu* – Chaga.

Distribution: Throughout India.

Taxonomic Description: Laticiferous shrubs, with fragrant whitish-Yellow flowers and mericarp fruits.

Medicinal Parts: Roots.

Chemical Constituents : Root - N-formylchone-morphine, N-methylchonemorphine, Japindine, Funtumafrine-C, Choneomorphin dihydrochloride, Glycerides of Lauric acid, Stearic acid, Arachidic acid, Myristic acid, palmitic acid, Oleic acid, Linolenic acid, β -Sitosterol, Taraxasterol.

Medicinal Properties:

मूर्वा सरा गुरूः स्वादुस्तिक्ता पित्तास्रमेहनुत् ।
त्रिदोषतृष्णा हृद्रोगकण्डूकुष्ठज्वरापहा ।। (भा. प्र.)

Mūrvā sarā gurūh ̣svāduṣtiktā pịttāsramehanuṭ
Tṛidosatṛṣṇāhṛdrogakaṇdūkuṣthajvarāpahā. (Bhā. Pṛa.)

Root - Sweet, Bitter, Astringent, Laxative, Thermogenic, Depurative, Carminative, Anthelmintic, Sudorific, Anodyne, Expectorant, Febrifuge.

Medicinal Uses: Root - Skin diseases, Leprosy, Scabies, Syphilis, Dyspepsia, Flatulence, Colic, Constipation, Helminthiasis, Cardiac debility, Diabetes, Jaundice, Cough, Bronchitis, Intermittent fevers.

Ethno- Medicinal Uses: Root - Stomach diseases (Kumar *et al.,* 1980).

CHROZOPHORA ROTTLERI (Geis.) A. Juss. ex Spreng. (EUPHORBIACEAE)

Syn.: *C. plicata* Hook. F.

Regional & Other Names : *English* – Turnsole; *Gujarati* - Kaloo-kharad; *Hindi* – Sonballi; *Marathi* – Suravarta; *Sanskrit* – Suryavarta.

Distribution: South, Western, Central & North India, Indo-China.

Taxonomic Description: Stellately hairy herbs, branched in the upper region. Leaves entire or 3-lobed, ovate to suborbicular. Sepals of female flowers triangular; stigma red. capsules purple.

Medicinal Parts: Seeds.

Chemical constituents: Root contains - Glycosides - 3,5,6,7,8-Pentamethoxyxanthone-1-O-rhamnosyl (1 $\rightarrow$ 6) - glucopyranosidej 3,5, 8 - trimethoxyxanthone - 1 - O - glucopyranoside and 2 - acetonyl-5-methyl-7-hydroxy-6C- glucophyranosyl - chromone - 2" - O- glucopyranoside.

Medicinal Properties: The plant has Corrosive Properties, Drastic, Emetic.

Medicinal Uses: Seeds - Cathartic.

Ethno- Medicinal Uses: Root - Cold (Sinha *et al.,* 2006); Cough (Singh & Pandey, 1980; Sinha *et al.,* 2006).

Seed - Purgative (Basak, 2006).

CICCA DISTICHIA L. (EUPHORBIACEAE)

Syn.: *C.acida* (L.) Merr.; *Phyllanthus distichus* Muell-Arg.; *P. acidus* Skeels.; *Averrhoea acida* L.

Regional & Other Names : *Bengali* – Noari; *English* – Country gooseberry; *Hindi* – Chalmeri; *Malayalam* - Nellipuli; *Sanskrit* – Lavali; *Tamil* - Sadagam; *Telugu* – Rachayusirika.

Distribution: Cultivated throughout India.

Taxonomic Description: Tree, with erect branches at top; branchlets slender. Leaves distichous thin. Obliquely ovate acute, base rounded, tip acute. Flowers very small in slender racemes. Fruits yellow-green, globose, fleshy, depressed both ends.

Medicinal Parts: Root, Fruits, Seeds.

Chemical Constituents: Root bark contains Tannin, Saponin, Gallic acid and a Crystalline substance, β-amyrin, Phyllanthol.

Medicinal Properties:

लवलीफलमस्त्रार्शः कफपित्तहरं गुरू ।
विशदं रोचनं रूक्षं स्वाद्वम्लं तुवरं रसे ।। (भा. प्र.)

Lavalīphalamaṣraṛṣaḥ kaphapiṭtaharaṃ guru
Viṣadaṃ rōcanaṃ ruksaṃ ṣvadvaṃlaṃ tuvaraṃ rase. (Bhā. Pṛa.)

Fruits - Appetiser, Astringent, Sour, Tonic.

Medicinal Uses: Fruits - Biliousness, Blood purifier Bronchitis, Constipation, Piles, Vomiting.

Ethno- Medicinal Uses: Root/Leaf/Fruits - Snake bite (Mishra & Sahu, 1984).

CITRUS AURANTIFOLIA (Christm.) Swingle (RUTACEAE)

Syn.: *C. aurantinum* var. *bigaradia* Hook f.

Regional & Other Names : *English* – Sour orange; *Hindi* – Khatta; *Kannada* – Bijapura; *Malayalam* - Vatukappulinarakam; *Sanskrit* – Brhat jambirah; *Tamil* – Narattai; *Telugu* – Bijauram.

Distribution: Throughout India.

Taxonomic Description: Medium sized shrubs or trees, with unifoliate leaves and fragrant white flowers.

Medicinal Parts: Fruits.

Chemical Constituents: Fruit juice yields 7-10% Citric acid, Phosphoric acid, Sugar. Fruit rinds yield a Volatile oil, a bitter Glucoside & Hesperidin.

Medicinal Properties:

बृहत् जम्बीरकं चाम्लं तुवरं तिक्तकं सरम् ।
उष्णं पित्तकफघ्नं च पाचनं परिकीर्त्तितम् ।।
ये गुणा लघुजम्बीरे ते वृद्धे सन्ति चाखिलाः ।। (शा. नि.)

Bṛhaṭ jaṃbīrakaṃ cāṃlaṃ tuvaraṃ tiktakaṃ saraṃ
Uṣnaṃ piṭtakaphaghnaṃ ca pācanaṃ parikīṛttitaṃ
Ye guṇā laghujaṃbīre vṛddhe saṇti cākhilaḥ. (Ṣa. Ni.)

Fruits – Anthelmintic, Appetiser, Astringent, Bitter, Digestive, Laxative, Sour, Stomachic, Thermogenic.

Medicinal Uses: Fruits - Cough, Bronchitis, Dyspepsia, Nausea, Flatulence, Colic, Helminthiasis, Scabies, Anaemia.

Ethno- Medicinal Uses: Fruits - Diarrhoea (Jain, 1965); Loose motion (Jain 1963).

Leaves - Headache, Fever (Deokule, 2006).

Fruit juice - Stomach disorders (Deokule, 2006).

CLEMATIS TRILOBA Heyne ex Roth. (RANUNCULACEAE)

Regional & Other Names : *Gujarati* - Morvel; *Hindi* – Churahar; *Marathi* – Moravela; *Sanskrit* – Morata.

Distribution: Bombay presidency; Konkan & Deccan, Western Ghats.

Taxonomic Description: An extensive climber, with opposite leaves and white flowers in axillary corymbose panicles.

Medicinal Parts: Roots.

Chemical Constituents: Anemonin, Alcoholic extract of plant is hypotensive and spamolytic.

Medicinal Properties:

मोरटा तुवरा तिक्ता स्वाद्विचोष्णा गुरुः स्मृता ।
पाककाले तु कटुका सारकाच त्रिदोषहा ।। (नि. र.)

Moratā tuvarā tiktā ṣvādvicoṣṇā gurūḥ ṣmṛtā
Pākakāle tu katukā sārakā ca ṭridosahā. (Ni. Ra.)

Leaves - Acrid, Alterative, Astringent, Bitter, Hot, Laxative, Sedative, Stimulant, Sweet.

Medicinal Uses: Leaves - Blood diseases, Syphilis, Scrofula, Leprosy, Chronic fevers, Heart troubles, Boils, Itching.

Ethno- Medicinal Uses: Boils, Itches (Janardhanan, 1963).

COLCHICUM LUTEUM Baker. (LILIACEAE)

Regional & Other Names : *English* – Golden collyrium; *Hindi* – Hirantutiya; *Sanskrit* – Hiranyaththa.

Distribution: Western temperate Himalayas from Kangra & Kashmir to Chamba.

Taxonomic Description: Corm bearing with herb; Leaves linear-oblong or oblanceolate, appearing with the flowers and short at the time of flowering; Flowers Yellow; Fruits capsules.

Medicinal Parts: Corm, Seeds.

Chemical Constituents: Corm - Colchicine, Starch.

Seed - Colchicine, Sugar, Fixed oil.

Medicinal Properties:

सुरञ्जानं तु कटुकं तिक्तं वीर्योष्णमेव च ।
कफवातहरं रक्तशोधनमामनाशनम् ।। (द्र. गु. वि.)

Suranjanaṃ tu katukaṃ tiktaṃ viryōsanamev ca
Kaphavātaharaṃ raktaśhodhanmāmnāśanaṃ. (Dra. Gu. Vi.)

Corm - Alterative, anti-Inflammatory, Anti-rheumatic, Aphrodisiac, Carminative.

Medicinal Uses: Corm - Diseases of the Spleen, Liver disorders.

Ethno- Medicinal Uses: Corms - Piles, Headache (Singh, 1996); Stomach problems, Gout, Joint pain (Ballabh & Chaurasia, 2006).

Whole plant - Rheumatism & Skin diseases (Kapur & Nanda, 1992).

COLEUS VETTIVEROIDES K.C. Jacob. (LAMIACEAE)

Syn.: *Plectranthus vettiveroides* (Jacob.) Singh & Sharma.

Regional & Other Names : *Hindi* – Valak; *Kannada* – Muchivala; *Malayalam* - Iruveli; *Sanskrit* – Valakam; *Tamil* – Kuruver; *Telugu* – Vettiveru.

Distribution: Cultivated throughout India.

Taxonomic Description: Small branched, succulent, aromatic herb with quadrangular stems & branches and deep straw coloured aromatic roots; Leaves broadly ovate with dentate morgines, glandular hair; Flowers blue; Fruits nutlets.

Medicinal Parts: Whole plant.

Medicinal Properties:

बालकं शीतलं रूक्षं लघु दीपनपाचनम् ।
हल्लासारूचिवीसर्पहृद्रोगामतिसारजित् ।। (भा. प्र.)

Bālakaṃ śītalaṃ rūksaṃ laghu dīpanapācanaṃ
Hṛḷlāsārucivīsaṛpahrdrogāmatisārajiṭ. (Bhā. Pṛa.)

The plant is Antipyretic, Bitter, Cooling, Diuretic.

Medicinal Uses: Burning sensation, diarrhoea, fever, hyperdipsia, leprosy, leucoderma, skin diseases, strangury, vomiting.

COMMELINA BENGHALENSIS L. (COMMELINACEAE)

Regional & Other Names : *Bengali* – Kanisira; *English* – Day flower; *Hindi* – Kankauoa; *Marathi* – Kena; *Sanskrit* – Kanchata; *Tamil* – Kanangakarai; *Telugu* – Nirukassuvu.

Distribution: Throughout India.

Taxonomic Description: Diffusely branched, deumbent herbs, with elliptic-ovate leaves. Flowers blue; spathes sessile or subsessile. Capsules 3-gonous; seeds pitted.

Medicinal Parts: Whole plant.

Chemical Constituents: The flower pigment contains delphinidin diglucoside as the main constituent and P-cumaric acid and awobanol.

Medicinal Properties: Whole plant – Bitter, Demulcent, Emollient, Laxative, Refringent.

Medicinal Uses: Whole plant - Leprosy.

Ethno- Medicinal Uses: Leaf - Diarrhoea (Saxena & Vyas, 1983); scorpion bite (Tribedi *et al.,* 1985); wounds (Kumar *et al.,* 1987); Scabies (Bondya *et al.,* 2006).

Root- Fever, liver complaint (Rajwar, 1983).

Leaves & Stem - Skin irritation (Arinathan *et al.,* 2003).

Whole plant - Scorpion sting (Sur *et al.,* 1992); Pimples (Kiurba *et al.,* 2006); Breast sores & Pimples (Rosakutty *et al.,* 2003); Jaundice (Bora, 2003); Better voice (Arinathan *et al.,* 2006).

COMMIPHORA MYRRHA (Nees.) Engl. (BURSERACEAE)

Regional & Other Names : *English* – Myrrh; *Hindi* – Bol; *Kannada* – Bola; *Malayalam* - Narumpasamaram; *Sanskrit* – Bolah; *Tamil* – Vellaippapolam; *Telugu* – Balimtra-Polam.

Distribution: Indigenous to North-East Africa especially Somalia Island.

Taxonomic Description: A small tree or Shrub.

Medicinal Parts: Gum.

Medicinal Properties:

बोलं तु कटुतिक्तोष्णं कषायं रक्तदोषनुत् ।
कफपित्तामयान् हन्ति प्रदरादिरूजापहम् ।। (रा. नि.)

Bolaṃ tu katutiktoṣṇaṃ kasāyaṃ raktadosanuṭ
Kaphapiṭtāmayāṇ haṇti ṗradarādirujāpahaṃ. (Rā. Ni.)

Gum - Acrid, Anthelmintic, Anti-Inflammatory, Antiseptic, Aphrodisiac, Astringent, Bitter, Carminative, Depurative, Digestive, Diuretic, Expectorant, Stimulant, Sudorific, Thermogenic, Tonic.

Medicinal Uses: Gum - Amenorrhoea, Asthma, Bronchitis, Dysmeno-rrhoea, Dyspepsia, Helminthiasis, Inflammation, Sciatica, Skin diseases, Stomatitis, Ulcers, Wounds.

CONVOLVULUS ARVENSIS L. (CONVOLVULACEAE)

Regional & Other Names : *Bengali* – Gondal; *English* – Dear's foot, Bind weed; *Gujarati* - Nari; *Hindi* – Harin padi; *Marathi* – Chanduel; *Sanskrit* – Rajbali; *Tamil* – Hiranpadi.

Distribution: Throughout India.

Taxonomic Description: Creeping or twining, perennial herbs, Leaves sagittate-hastate at base, Flowers pale pink,

Medicinal Parts: Whole plant.

Chemical Constituents: Roots - Cuscohygrine, Coumarins.

Aerial parts - N-alkanes, N-alkanols, □-amyrin & Sterols.

Medicinal Properties:

प्रसारणी गुरूर्वृष्या बलसन्धानकृत्सरा ।
वीर्योष्णा वातहृत्तिक्ता वातरक्तकफापहा ।। (भा. प्र.)

Prasāranī gururvrusyā balasaṇdhānkriṭsarā
Viryōsanā vāthrttiktā vātaraktakaphāpahā (Bha. Pra.)

Plant extract - Anticonvulsant, Hypotensive, Mascarinic & Nicotinic.

Medicinal Uses: Leaves - Boils.

Ethno- Medicinal Uses: Root - Purgative (Das *et al.*, 1983; Sharma *et al.*, 1979; Deokule & Magdum, 1992; Sinha *et al.*, 2006; Deokule, 2006).

CORCHORUS OLITORIUS L. (TILIACEAE)

Regional & Other Names : *Bengali* – Mithapat; *English* – Jew's mallow; *Hindi* – Koshta; *Marathi* – Motichunch; *Sanskrit* – Pattshak; *Tamil* – Peratti; *Telugu* – Parinta.

Distribution: Throughout India.

Taxonomic Description: Annual herbs, with ovate-lanceolate leaves & yellow flowers. Capsules 5-valved, 10 ribbed, with entire beak.

Medicinal Parts: Leaves, Seeds.

Chemical Constituents: Fibers - Cellulose, Lignin, Nitrogen, Fat, Wax.

Seeds - Crude protein, Crude fibre, Free invert sugar, Minerals; Fixed oil; Fatty acids; Oleic & Linoleic acid, Arachidic acid, Palmitic, Stearic acid; Crystalline Substance-Corchoritin.

Medicinal Properties:

पट्टशाकस्तु नाडीको नाडीशाकश्च स स्मृतः ।
नाडीको रक्तपित्तघ्नो विष्टम्भो वातकोपनः ।। (भा. प्र.)

Paṭtashākaṣtu nādīko nādīṣhākaṣya sa .smṛitaḥ
Nādīko raktapịttaghno vịstaṃbho vātakopanaḥ (Bhā. Pṛa.)

Leaves - Acrid, Alexteric, Alterative, Astringent, Bitter, Demulcent, Diuretic, Hot, Tonic.

Seeds - Purgative.

Medicinal Uses: Leaves - Abdominal tumours, Ascites, Pain, Piles, Tumours.

Ethno- Medicinal Uses: Seed - Purgative (Shah *et al.*, 1983; Sebastian & Bhandari, 1984; Sinha *et al.*, 2006).

Whole plant - Stomachache (Jadeja *et al.*, 2006).

Leaf- Piles (Sebastian & Bhandari, 1984); Boils (Ganesan *et al.*, 2006); Fever, Amoebiasis (Tomar & Singh, 2006).

COSCINIUM FENESTRATUM (Gaertn.) Colebr. (MENISPERMACEAE)

Regional & Other Names : *English* – Tree turmeric; *Hindi* – Jhade-halde; *Kannada* – Maradasina; *Malayalam* - Maramannal; *Marathi* – Jhade-halade; *Sanskrit* – Darvi; *Tamil* – Mancalkoti; *Telugu* – Manu pasupu.

Distribution: Western ghats, in Tamin Nadu & Kerala.

Taxonomic Description: Woody climbing shrubs, with Yellow coloured unisexual flowers and Drupes fruits.

Medicinal Parts: Stem.

Chemical Constituents: Stem contains Berberine and Saponin.

Medicinal Properties-

तिक्ता दारूहरिद्रा तु कटुष्णा व्रणमेहनुत् ।
कण्डूविसर्पत्वग्दोषविषकर्णाक्षिदोषहा।। (रा. नि.)

Tiktā dārūharidṛā tu katuṣṇā vṛanamehanuṭ.
Kaṇdūvisaṛpaṭvagdosavisakaṛṇāksidosahā. (Rā. Ni.)

Stem - Anodyne, Anti-Inflammatory, Antiseptic, Bitter, Depurative, Febrifuge, Stomachic, Sudorific, Thermogenic, Tonic.

Medicinal Uses- Stem - Abdominal disorders, Diabetes, Fever, General debility, Inflammation, Jaundice, Skin diseases, Ulcers, Wounds.

Ethno- Medicinal Uses: Stem & Root bark -Jaundice (Udayan *et al.*, 2006).

CRINUM DEFIXUM Ker.-Gwal. (AMARYLLADACEAE)

Regional & Other Names : *Bengali* – Sukdarshan; *Gujarati* - Nagrikand; *Sanskrit* –Vishamandala; *Telugu* – Kesarchettu.

Distribution: Swampy river-banks throughout India.

Taxonomic Description: A perennial herb with avoid bulb; leaves linear, subfleshy, slightly concave, apex acute; scape 30-50 cm. long; bracts spathaceous, membranous; flowers white; fruit sub-globose, beaked; seeds 1-2, rogouse.

Medicinal Parts: Bulb, Leaves.

Chemical Constituents: Bulb - Lycorine.

Medicinal Properties:

दध्याली स्वादुतिक्तोष्णा कफशोथास्त्रवातजित् ।। (कै. नि.)

Dadhyālī ṣvādutiktoṣṇā kaphaśothāṣravātajiṭ. (Kaī. Ni.)

Bulb - Diaphoretic, Emetic, Emollient, Nauseant.

Medicinal Uses: Bulb - Burns, Carbuncle, Witlow.

Ethno- Medicinal Uses: Bulb - Abdominal diseases (Venugopal & Raghunathan, 1982); Stomach pain (Sinha *et al.,* 2006); Bodyache (Singh & Maheshwari, 1983).

Leaf - Earache (Lal & Yadav, 1983).

CROTON OBLONGIFOLIUS Roxb. (EUPHORBIACEAE)

Syn.: *C.roxburghii* Balak.

Regional & Other Names : *Bengali* – Baragach; *Hindi* – Chucka; *Malyalam* - Kote putol; *Marathi* – Gunsur; *Sanskrit* – Bhutankusa, Nagadanti; *Tamil* – Milgunari; *Telugu* – Bhutalabhairi.

Distribution: Bengal, Chota Nagpur, Central Provinces, Burma, Western Peninsula, Ceylon.

Taxonomic Description: Middle sized trees, with Pale Yellowish-Green flowers and Capsule fruits.

Medicinal Parts: Root bark, Leaves, Fruits, Seeds.

Chemical Constituents: β-Sitosterol, a Diterpene Alcohol, Oblongifoliol, Deoxyobiongifoliol, Acetylaleuritolic acid, Quercetin & Procyanidin.

Medicinal Properties:

नागदन्ती सरा चोष्णा कटुका क्रिमिनाशिनी ।
शूलकुष्ठामदोषघ्नी मेदोरोगाश्मरीहरा ।।
मुखरोगहरा चान्ये गुणास्तु लघुदन्तिवत् ।। (नि. र.)

Nāgadaṇtī sarā coṣṇā katukā kṛimināṣinī
Śulakuṣthāmadosaghnī medorogāṣmrīharā
Mukharōgaharā cāṇye guṇāstu laghudaṇtivaṭ. (Ni. Ra.)

Bark / Root - Alterative.

Seed / Fruits - Purgative.

Medicinal Uses: Root - Constipation.

Root bark - Fever, Inflammation, Liver diseases, Sprains.

Ethno- Medicinal Uses: Root - Antifertility (Goel *et al.,* 1984); Constipation, Diarrhoea, Dysentery (Mudgal & Pal, 1980; Maheshwari *et al.,* 1986; Goel *et al.,* 1984; Pal & Srivastava, 1976); Rheumatism (Jain & De, 1966); Snake bite (Mishra & Sahu, 1984;, Goel *et al.,* 1984); Sore throat (Srivastava & Varma, 1981); Vomit (Maheshwari *et al.,* 1986); Sprain (Chaudhuri *et al.,* 1975).

Bark - Constipation, Diarrhoea, Dysentery (Saxena *et al.,* 1981); Snake bite (Mishra & Sahu, 1984).

Fruits - Snake bite (Mishra & Sahu, 1984)

Leaf - Wounds (Jain & De, 1966).

CUCUMIS MELO var. *UTILISSIMUS* (Roxb.) Duthie & Fuller. (CUCURBITACEAE)

Regional & Other Names : *Bengali* – Kakur; *English* – Snake cucumber; *Hindi* – Kakkri; *Malayalam* - Kakkarikka; *Sanskrit* – Karkati; *Tamil* – Kakkarikkay; *Telugu* – Kharbujadosa.

Distribution: Cultivated throughout North India.

Taxonomic Description: A creeping annual. Flowers Yellow. Fruits Oval or Cylindrical to elongate, straight or curved, dark Green to nearly White, Bright, Orange when ripe.

Medicinal Parts: Fruits, Seeds.

Chemical Constituents: Seeds contain much farinaceous matter blended with a large proportion of a Mild oil.

Medicinal Properties:

कर्कटी शीतला रूक्षा ग्राहिणी मधुरा गुरूः ।
रूच्या पित्तहरा सामा पक्वा तृष्णाग्निपित्तकृत्त ।। (भा. प्र.)

Karkatī śītalā rūksā grāhiṇī madhurā gurūḥ
Rūcyā pittaharā sāmā pakvā tṛsnāgnipittakṛt. (Bhā. Pra.)

Fruits - Acrid, Aphrodisiac, Appetiser, Cooling, Diuretic, Laxative, Sweet, Tonic.

Seeds - Cooling, Diuretic, Nutritive.

Medicinal Uses: Fruits - Insanity, Ascites, Ophthalmia, Dyspepsia, General debility.

Seeds - Ophthalmia, Hepatopathy, Nephropathy, Bronchitis, Chronic fever, Burning sensation.

Ethno- Medicinal Uses: Seed - Cooling (Dar *et al.,* 1984).

Fruits - Gout (Jadeja *et al.,* 2006).

CUCUMIS SATIVUS L. (CUCURBITACEAE)

Regional & Other Names : *Bengali* – Khira; *English* – Cucumber; *Hindi* – Khira; *Kannada* – Mullu savte; *Malayalam* - Vellari; *Marathi* – Khira; *Sanskrit* – Trapusah; *Tamil* – Vellarikkay; *Telugu* – Dosakay.

Distribution: Cultivated throughout India.

Taxonomic Description: A scabrous, trailing or climber annual. Leaves 3 to 5 lobed, cordate, very rough. Flowers Yellow; Male ones in axillary clusters and females solitary or fascicled. Fruits ovate or oblong, very variable in size. Seeds numerous, White.

Medicinal Parts : Fruits, Seeds.

Chemical Constituents: Fruits contain - Erepsin enzyme, Vitamin B & C; proteolytic enzymes, Ascorbic acid, Oxidase, Succinic & Malic dehydrogenases, rutin, Palmitic, Stearic, Oleic acids, their Glycerides. Sterols & Squalene.

Seeds contain - Sitosterol, Cucurbitaside.

Leaf contains - Ferrodoxin, free Cucurbitasides B & C.

Medicinal Properties:

स्यात् त्रपुसीफलं रूच्यं मधुरं शिशिरं गुरू ।
भ्रमपित्तविदाहार्त्तिवान्तिहृद्‌बहुमूत्रदम् ।। (रा. नि.)

Ṣyāṭ ṭrapusīphalaṃ rūcyaṃ madhuraṃ śiśiraṃ gurū
Bhṛamapiṭtavidāhaṛttivaṇtihṛdbahumūṭradaṃ. (Rā. Ni.)

Fruits - Diuretic, Haemostatic, Refrigerant, Sweet, Tonic.

Seeds – Antipyretic, Astringent, Diuretic, Nurtitive, Purgative, Refrigerant, Sweet, Tonic.

Medicinal Uses: Fruits - Bronchitis, Burning sensation, Fever, General debility, Haemorrhages, Insomnia, Jaundice, Strangury.

Seeds - Burning sensation, Constipation, General debility, Intermittent fever, Renal calculus, Strangury.

Ethno- Medicinal Uses: Leaf - Fever (Maheshwari & Singh,1984).

Fruits - Whooping cough (Megoneitso, Rao, 1983).

Seed - Facial cosmetic (Rajendra & Aswal, 2003).

CUCUMIS TRIGONUS Roxb. (CUCURBITACEAE)

Syn.: *C. callosus* Cogn.

Regional & Other Names : *Bengali* – Kundruki; *English* – Bitter gourd; *Gujarati* - Kothiban; *Hindi* – Jangli Indrayan; *Kannada* – Karanti; *Malayalam* - Pekkummatti; *Marathi* – Karit; *Sanskrit* – Indravaruni; *Tamil* – Kattutummatti; *Telugu* – Adavi pucca.

Distribution: Throughout India.

Taxonomic Description: Perennial tendrillar herbs, with palmately five-lobed leaves and Yellow unisexual flowers.

Medicinal Parts: Roots, Fruits, Seeds.

Chemical Constituents: Seed oil - Linoleic acid, Oleic acid, Stearic aicd, Arachidic acid, Behenic acid, Myristic acid, Palmitic acid.

Fruits - Cucurbitacin-B, D & Q, Chondrillasterol, β-Stiosterol.

Medicinal Properties:

इन्द्रवारूणिका तिक्ता कटुः शीता च स्वेदिनी ।
गुल्मपित्तोदरश्लेष्मकृमिकुष्ठज्वरापहा ।। (रा. नि.)

Indravāruṇikā tiktā katuḥ śītā ca svedinī
Guḷmapiṭtodarśleṣmakṛmikuṣthajvarāpahā. (Rā. Ni.)

Fruit pulp - Acrid, Anthelmintic, Appetiser, Bitter, Expectorant, Febrifuge, Purgative, Stomachic, Thermogenic.

Seeds – Astringent, Cooling.

Medicinal Uses- Fruit pulp - Anaemia, Bronchitis, Constipation.

Cough, Diabetes, Fever, Flatulence, Helminthiasis, Jaundice, Leprosy.

Ethno- Medicinal Uses: Fruits - Skin diseases (Sebastian & Bhandari, 1984), Wounds (Upadhye *et al.,* 1986).

CUCURBITA MAXIMA Duchesne ex Lam. (CUCURBITACEAE)

Regional & Other Names : *Bengali* – Kumhara; *English* – Red gourd pumpkin; *Gujarati* - Pilukahalu; *Hindi* – Kaddu; *Kannada* – Gumbaloo; *Malayalam* - Mattan; *Marathi* – Kohola; *Sanskrit* – Pitakusmandah; *Tamil* – Pusanikkay; *Telugu* – Gummadi.

Distribution: Cultivated throughout India.

Taxonomic Description: Leaves circular to reniform in outline. Flowers Yellow. Peduncle short, spongy.

Medicinal Parts: Fruits, Seeds.

Chemical Constituents: Leaves - Calcium, Magnesium, Iron, Zinc, Copper.

Seed - Euglobulin, □-Spinasterol, Linoleic acid, Oleic acid, Linolenic acid, β-tocopherol, β-Sitosterol, Cholesterol, Ergosterol, Campesterol, Codisterol, Clerosterol, Isofucosterol.

Medicinal Properties:

अपरं पीतकूश्माण्डं गुरू पित्तकरं परम् ।
अग्निमान्द्यकरं स्वादु श्लेष्मध्नं वातकोपनम् ।। (आ. सं.)

Aparaṃ pītakūṣmāṇḍaṃ gurū piṭtakaraṃ paraṃ
Agnimāṇdyakaraṃ ṣvādu śleṣmaghnaṃ vātakopanaṃ. (A. Sam.)

Fruits - Diuretic, Emollient, Refrigerant, Sedative, Sweet, Tonic.

Seeds - Anthelmintic, Diuretic, Nervine tonic.

Medicinal Uses: Fruits - Boils, Burns, Inflammation, Migraine, Neuralgia, Scalds.

Seeds - Nervous debility, Strangury.

Ethno- Medicinal Uses: Seed - Intestinal worms (Kamble & Pradhan, 1980).

Fruit pulp - Snake bite (Desmukh & Rothe, 2003).

Dried stalk - Cold & Cough (Pereira, 2006).

Fruit - Poultice is applied for inflammed parts (Deokule, 2006).

CYAMOPSIS TETRAGONOLOBA (L.) Taub. (FABACEAE)

Syn.: *C. psoralioides* (L.) DC.

Regional & Other Names : *English* – Cluster bean; *Gujarati* –Gawaru ni phali; *Hindi* – Govar; *Kannada* – Gorikayi; *Malayalam* - Kottavara; *Marathi* – Gawar; *Sanskrit* – Ksudrasimbi; *Tamil* – Kottavarai; *Telugu* – Gorucikkudu.

Distribution: Cultivated throughout India.

Taxonomic Description: An erect robust annual. Leaflets 3, ovate-elliptic, Coarsely toothed. Flowers Pinkish-purple, in axillary racemes. Pods erect. Seeds dull, Compressed.

Medicinal Parts: Pods.

Chemical Constituents: Pods - Methionine, Globulin, Gallic acid, Caffic acid, Chlorogenic acid, Ellagic acid, Gallotannins, Kaempferol.

Seeds - Gallic acid, Galactose, Saponins.

Medicinal Properties:

बाकुचिकाशिम्बि रूक्षा वातला मधुरा गुरूः ।
सरा कफकरी चाग्निदीपनी पित्तनाशिनी ।।
पत्रमस्य निशानध्यघ्नं पित्तनाशकरं परम् ।। (शा. नि.)

Bākucikāśimbi ruksā vātalā madhurā guruḥ
Sarā kaphakarī cāgnidīpanī pittanāśinī
Paṭramaṣya niśāndhyaghnaṃ piṭtanāṣakaraṃ paraṃ. (Sā. Ni.)

Pods - Appetiser, Cooling, Digestive, Laxative, Sweet, Tonic.

Medicinal Uses- Pods – Agalactia, Anorexia, Constipation, Dyspepsia.

Ethno- Medicinal Uses: Seed - Sprain, Swelling (Shekhawat & Sunil Anand, 1984; Sebastian & Bhandari, 1984).

CYATHULA PROSTRATA (L.) Blume. (AMARANTHACEAE)

Syn.: *Desmochaeta prostrata* (L.) DC.

Regional & Other Names : *English* – Small prickly chaff-flower plant; *Hindi* – Lal cicimda; *Kannada* – Uttarani; *Malayalam* - Cerukatalati; *Sanskrit* – Raktapamargah; *Tamil* – Cirukatalati; *Telugu* – Antisa.

Distribution: Throughout India.

Taxonomic Description: Annual or perennial herbs, with reddish stem-violet stem & branches and Violet flowers.

Medicinal Parts: Roots.

Medicinal Properties:

रक्तोऽपामार्गकः शीतः कटुकः कफवातनुत् ।
व्रणकण्डूविषघ्नश्च संग्राही वान्तिकृत् परः ।। (रा. नि.)

Raktoapāmāṛgakaḥ śitaḥ katukaḥ kaphavātanuṭ
Vṛanakaṇdūvisaghnaṣca saṃgṛāhi vāṇtikṛṭ paraḥ. (Rā. Ni.)

Root - Acrid, Alexipharmic, Constipating, Cooling, Emetic, Expectorant, Vulnerary.

Medicinal Uses: Root - Abdominal disorders, Bronchitis, Cough, Diarrhoea, Dyspepesia, Pruritus, Skin diseases, Ulcers, Wounds.

Ethno- Medicinal Uses: Px - Skin diseases (Das *et al.*, 1983).

Root - Tuberculosis (Chaudhury & Neogi, 2003); Dysentery (Das *et al.*, 2006).

Leaf - Cuts, Wounds (Chaudhury & Neogi, 2003).

CYCLEA PELTATA (Lam.) Hook. F. & Thoms. (MENISPERMACEAE)

Syn.: *C. burmanni* Hook. f. & Thoms.

Regional & Other Names : *English* – Pata root; *Hindi* – Path; *Kannada* – Padvali; *Malayalam* - Patavalli; *Marathi* – Patha; *Sanskrit* – Patha; *Tamil* – Patakkilanku; *Telugu* – Patha.

Distribution: Throughout India.

Taxonomic Description: Twining shrubs, with Greenish small flowers and drupes fruits.

Medicinal Parts : Root, Leaves.

Chemical Constituents: Root - Telradirine, Phaeanthine, Uronic acid, Fangchinoline, □-tetranoandrine, Cycleapeltine, Cycleadrine, Cycleacurine, Cycleanorine, Cycleanomine Chloride, Chondocurine, Curine, Homoaromaline, Isotetradrine, Berbamine, Fangchinoline, Cycleanine.

Medicinal Properties:

विषघ्नी सर्वशूलघ्नी गुल्मघ्नी वातपित्तजित् ।
पाठा ज्वरप्रशमनी रक्तातीसारनाशनी ।। (म. नि.)

Visaghnī saṛvaśūlaghnī gulmaghnī vātapiṭtajiṭ.
Pāthā jvarapṛaśamanī raktātīsāranāśanī. (Ma. Ni.)

Roots - Acrid, Anthelmintic, Anti-inflammatory, Bitter, Carminative, Constipating, Depurative, Digestive, Diuretic, Febrifuge, Thermogenic, Tonic.

Leaves - Cooling, Ophthalmic.

Medicinal Uses: Root - Cough, Bronchitis, Dyspepsia, Colic, Diarrhoea, Dysentery, Dropsy, Leprosy, Skin diseases, Pruritus, Fever, Ulcers, Wounds, Vomiting, Cardiac disorders.

Leaves - Burning sensation.

Ethno- Medicinal Uses: Root - Small pox (Kumar *et al.*, 1980); Stomach pain (Shivakumar *et al.*, 2003; Rajan *et al.*, 2003; Ramachandran & Nair, 1981; Upadhye *et al.*, 1986; Udayan *et al.*, 2006); Wound infections (Udayan, *et al.*, 2006); Diarrhoea (Sasidharan & Augustine, 2006).

Leaf - Malaria fever, Typhoid fever (Maikhuri & Ramakrishanan, 1992).

CYPERUS SCARIOSUS R.Br. (CYPERACEAE)

Regional & Other Names : *Bengali* – Nagarmutha; *Gujarati* - Nagaramothya; *Hindi* –Nagarmutha; *Malayalam* - Korakizhana; *Marathi* – Lawala; *Sanskrit* – Bhadramusta; *Tamil* – Koraikkilangu; *Telugu* – Kolatungamuste.

Distribution: Bengal, Pegu.

Taxonomic Description: Glabrous, Stolon slender, clothed by scales, stems long, slender, triquetrous. Leaves variable, sometimes absent umbel thin slender contracted, ray slender. Spikelets linear, pale straw coloured.

Medicinal Parts: Tubers & Oil.

Chemical Constituents: Essential oil from tuber-Sesquiterpenoid, cyperenone; Cyperenol and Patchoulenol, Isopatchoulenone. Rotundene and Rotundenol from oil; in addition to (-) β-selinene, Isopathoula-3, 5-diene. Leaves glycoside of leptosidin.

Medicinal Properties:

भद्रमुस्ता तु तुवरा शीता तिक्ता च पाचका ।
कटुग्निदीपनी ग्राही चाम्लपित्तकफापहा ।। (नि. र.)

Bhaḍramuṣtā tu tuvarā śītā tiktā ca pācakā
Katugnidīpanī g̣rāhī cāṃlapiṭtakaphāpahā. (Ni. Ra.)

Tubers - Aromatic, Astringent, Diaphoretic, Diuretic, Stomachic.

Essential oil - Hypotensive & CNS stimulant.

Medicinal Uses: Tuber - Diarrhoea, Gonorrhoea, Syphilis.

Ethno- Medicinal Uses: Tuber - Fever (Singh & Kumar, 2003); Snake bite, Cholera (Maheshwari *et al.*, 1986).

DELPHINIUM DENUDATUM Wall. ex Hook. f. & Thoms. (RANUNCULACEAE)

Regional & Other Names : *Gujarati* - Nirvishi; *Hindi* – Nirbisi; *Kannada* – Nirvishi; *Malayalam* - Nirvasi; *Marathi* – Nirvishi; *Sanskrit* – Nirvisa; *Tamil* – Nirbasi .

Distribution: In the Himalayas, upto 3,600 m.

Taxonomic Description: Annual glabrous herbs, with pinnately lobed leaves and Blue flowers.

Medicinal Parts: Roots.

Chemical Constituents: Root - Sterols as Campesterol, Stigmasterol, Sitosterol, Cholesterol, D5-avenasterol alkaloids as denudatine, Condelphine, Talatizidine, Iso-talatizine, Fatty acids as Capric acid, Myristic acid, Palmitic acid, Palmitioleic acid, Stearic acid, Oleic acid, Linoleic acid.

Medicinal Properties:

निर्विषा कटुका सोष्णा कफवातास्त्रदोषनुत् ।
अनेकविषदोषघ्नी व्रणसंरोपणी च सा ।। (रा. नि.)

Nirvisā katukā sosṇā kaphavātāṣradosanut
Anekavisadosaghni vranasamropaṇi ca sā. (Rā. Ni.)

Roots - Acrid, Bitter, Anodyne, Anthelmintic, Anti-inflammatory, Aphrodisiac, Carminative, Depurative, Digestive, Diuretic, Febrifuge, Stimulant, Thermogenic, Vulnerary.

Medicinal Uses: Roots - Amenorrhoea, Asthma, Cough, Dysmenorrhoea, Dyspepsia, Fever, Flatulence, Haemorrhoids, Inflammation, Insanity, Jaundice, Leprosy, Obesity, Skin diseases, Strangury, Ulcers.

Ethno- Medicinal Uses: Root - Abdominal pain (Uniyal, 1968); Anthelmintic (Nautiyal, 1981); Respiratory complaint, Toothache (Nautiyal, 1981); Ulcers (Shah & Joshi, 1981); Toothache, Kidney & Gall bladder stones (Chhetri, 2005); Hepatic ulcer & impotency (Arya & Agarwal, 2006).

Whole plant - Ulcer, Urinary disorders, Fever, Calculus (Kapur & Nanda, 1992); Fever (Kapur *et al.,* 1992).

Fruits - Cough (Verma & Chauhan, 2006).

DENDROCALAMUS STRICTUS (Roxb.) Nees. (POACEAE)

Regional & Other Names : *Bengali* – Halpa; *English* – Solid bamboo; *Gujarati* – Nakor vans; *Hindi* – Bans; *Kannada* – Kiribidiru; *Malayalam* - Arinkantam; *Marathi* – Bans; *Sanskrit* – Vansha; *Tamil* – Karanai; *Telugu* – Veduru.

Distribution: Throughout India.

Taxonomic Description: Arborescent or shrubby bamboos, with smooth culms. Leaves with a petiole like base & articulated with sheath. Spikelets in dense clusters along the panicle branches. Rachilla internodes very short, not articulating. Fruits subglobose.

Medicinal Parts: Clum.

Chemical Constituents: Silicious matter.

Medicinal Properties: Clum - Tonic, Astringent, Cooling.

Ethno- Medicinal Uses: Root - Antifertility (Sharma, 1981); Fever (Mishra & Billore, 1983).

Bark - Injuries (Joshi *et al.,* 1980).

Siliceous matter - Tonic (Deokule, 2006; Deokule & Magdum, 1992).

Shoot - Tonic (Roy & Chaturvedi, 1986).

Leaf - Cough (Sebastian & Bhandari, 1984); Cough, Affections (Kapur et al., 1992); Easy deliver (Chhetri, 2005); Sore throat (Jadhav, 2006).

DESMODIUM TRIFLORUM (L.) DC. (FABACEAE)

Regional & Other Names : *Bengali* – Kodaliya; *Gujarati* - Jhinopandhio; *Hindi* – Kudaliya; *Kannada* – Muntamunda; *Malayalam* - Nilamparanta; *Marathi* – Jungli methi; *Sanskrit* – Tripadi; *Tamil* – Sirupullati; *Telugu* – Muntamandu.

Distribution: Throughout India.

Taxonomic Description: A much branched, gregarious, wiry, trailing herbs, rooting at the nodes. Leaflets obovate, obcordate. Flowers Pink or Purplish, 1-3 in the axils of leaves. Pods curved, flat, indented on the lower edge, 3 to 5 jointed. Seeds Brown.

Medicinal Parts: Whole plant.

Chemical Constituents: Root / Leave / Stem - Trigonelline, Stachydrine, Betaine, Choline, Indole-3-acetic acid, Tyramine, Hypaphorin, 4β-Phenylethylamines.

Medicinal Properties:

हंसपादी गुरू: शीता हन्ति रक्तविषव्रणान् ।
विसर्पदाहातिसारलुताभूतादि रोगनुत् ।। (भा. प्र.)

Haṃsapādī guruḥ śītā haṇti raktavisaẏraṇāṇ
Visaṛpadāhātisāralutābhutādi roganụt. (Bhā. Pṛa.)

The plant is Acrid, Cooling, Expectorant, Sweet.

Medicinal Uses: The plant is useful in Bronchitis, Burning sensation, Cough, Dysentery, Erysipelas, Flatulence, Pruritus, Wounds.

Ethno- Medicinal Uses: Whole plant – Bodyache, Swelling (Goel *et al.*, 1984); Corns (Binu *et al.*, 2003); Scorpion sting (Singh *et al.*, 1992).

Leaf - Eruption (Sebastian & Bhandari, 1984); Diarrhoea (Singh & Pandey, 1980; Puri, 1983; Sinha *et al.*, 2006); Dysentery (Malhotra & Moorthy, 1973; Singh & Pandey, 1980; Sharma *et al.*, 1979); Toothache (Ramachandran & Nair, 1981; Sinha *et al.*, 2006); Wounds (Kapur *et al.*, 1992; Singh & Pandey, 1980; Sebastian & Bhandari, 1984; Rajwar, 1983; Bhalla *et al.*, 1992); Boils, Gout, Lumbago (Dwivedi & Pandey, 1992); Eruption (Kapur *et al.*, 1992); Breast pain (Sinha *et al.*, 2006).

DIGERA ARVENSIS (L.) DC. (AMARANTHACEAE)

Syn.: *D. muricata* (L.) Mart.

Regional & Other Names : *Bengali* – Gungatiya; *Gujarati* - Kanajaro; *Hindi* – Latmahuria; *Marathi* – Gitana; *Sanskrit* – Aranyavastuka; *Tamil* – Thoyyakeerai; *Telugu* – Chanchali kura.

Distribution: Throughout India.

Taxonomic Description: Erect or decumbent herbs; leaves ovate-lanceolate to oblong, entire or crenulate, acute, obtuse or abruptly acuminate. Flowers pink, in axillary, 10-30 cm. long racems; bacts oblong, persistent, 2 outer tepals 3-4 mm long, 5-7 nerved; 3 inner tepals shorter, narrow 1-2 nerved; fruits ovate-globose, with 2-fid apex, 1-seeded.

Medicinal Parts: Flowers and Seeds.

Chemical Constituents: Leaves - The Protein, Fat, Total mineral, Crude fibre, Carbohydrates. Calcium, Iron, Ascorbic acid and β -carotene.

Medicinal Properties- Acrid, Astringent,Cooling, Laxative, Sweetish.

Medicinal Uses: Flower / Seeds - Urinary discharges.

Ethno- Medicinal Uses: Leaf- Boils (Shekhawat & Sunil Anand, 1984).

DIGITALIS PURPUREA L. (SCROPHULARIACEAE)

Regional & Other Names : *English* – Foxglove; *Hindi*-Digitalis; *Sanskrit* – Tilapuspi.

Distribution: Cultivated in Kashmir, Darjeeling & Nilgiris.

Taxonomic Description: A tall binneal herbaceous plants upto a height of 1.8 m.

Medicinal Parts: Leaves

Chemical Constituents: Leaves contains - Approximately 30 Glucosides; 3 main Glucosides are Digitoxin, Gitoxin, Gitalin other important Glucosides are Digitoxoside, Glucogitaloxigenin-bis-digitoroside, Glucoeva-tromonoside, Glucolanadoxin, Verodoxin & Stropeside. The leaves also contain anthraquinones.

Medicinal Properties: Anti-inflammatory, Aphrodisiac, Cardio-tonic, Diuretic, Expectorant, Haemostatic, Heart Stimulant, Irritant.

Medicinal Uses: Asthma, Blood pressure, Cough, Dropsy, Dysmeno-rrhoea, Fever, Liver & Spleen diseases, Nervous disorders, Tuberculosis.

DIOSPYROS MELANOXYLON Roxb. (EBENACEAE)

Syn.: D. tupru Buch.-Ham.

Regional & Other Names : *Bengali* – Bhum; *English* – Coromandel ebony persimmon; *Gujarati* - Tamrug; *Hindi* – Tendu; *Kannada* – Abanasi; *Malayalam* - Kari; *Marathi* – Tendu; *Sanskrit* –Dirghpatraka; *Tamil* – Karundumbi; *Telugu* – Nallatumki.

Distribution: Throughout India.

Taxonomic Description: Deciduous trees, with sub-opposite & alternate, ovate-elliptic leaves. Male flowers in drooping cymes; anthers awnless. Fruits globose or ovoid.

Medicinal Parts: Bark, Leaves, Fruits.

Chemical Constituents: Bark - Tannins.

Medicinal Properties: Leaves - Diuretic, Carminative, Laxative, Styptic.

Bark - Astringent.

Fruits - Carminative, Laxative, Diuretic.

Medicinal Uses: Dried flowers - Urinary, Skin & Blood diseases.

Bark - Diarrhoea, Dyspepsia.

Ethno- Medicinal Uses: Seed - Diarrhoea (Satapathy & Brahman, 2003); Dysentery (Joshi, 1982); Diabetes (Ghosh, 2006).

Flower- Skin diseases (Malhotra & Moorthy, 1973).

Fruits - Urinary compliant (Bhatt & Sabnis, 1987).

Gum - Dysentery (Upadhyay & Chauhan, 2003).

Bark - Skin diseases, Scabies (Ravikumar & Sankar, 2003); Diarrhoea (Satapathy & Brahman, 2003; Dwivedi, 2003).

Unipe fruit - Colic, Intestinal pain (Misra, 1992).

Root - Diabetes (Ghosh, 2006).

Leaves - Gonorrhoea (Prakash & Singh, 2006); Diabetes (Ghosh, 2006).

DIOSPYROS MALABARICA Kostal. (EBENACEAE)

Syn.: *D. peregrina* auct. non Gruke; *D. embryopteris* Pers.

Regional & Other Names : *Bengali* – Gab; *English* –Gaub Persimmon; *Hindi*- Kalatendu; *Kannada* – Kusharta ; *Malayalam* -Vananji ; *Marathi* – Timburi; *Sanskrit* – Tinduka; *Tamil* – Kattatti; *Telugu* –Tinduki.

Distribution: Throughout India.

Taxonomic Description: A small or middle sized tree.

Medicinal Parts: Bark, Flowers, Fruits, Seed oil.

Chemical Constituents: Raw fruits contain tannin.

Medicinal Properties: Unripe Fruit- Acrid, Astringent, Bitter, Indigestible.

Bark- Febrifuge.

Medicinal Uses: Unripe fruit- Ulcers.

Ripe fruit- Biliousness, Diseases of blood, urinary losses, Stone in the urinary tract.

Seed oil- Diarrhoea, Dysentery.

Flowers & Fruits- Hiccough in children.

Bark- Dysentery.

Wood- Biliousness.

Ethno- Medicinal Uses: Fruits - Abortifacient (Tarafder, 1983); Blisters mouth; Diarrhoea (Jain 1965); Dysentery; Tonic (Roy & Chaturvedi, 1986).

Stem bark - Abortifacient (Tarafder, 1983).

EMILIA SONCHIFOLIA (L.) DC. (ASTERACEAE)

Regional & Other Names : *Bengali* – Sadhimodi; *Hindi* – Hirankhuri; *Malayalam* - Muyalccevi; *Sanskrit* – Sasasrutih; *Tamil* – Muyalccevi.

Distribution: Throughout India.

Taxonomic Description: Variously branched procumbent annual herbs, with Purple flowers and Achene fruits.

Medicinal Parts: Whole plant.

Chemical Constituents: Aerial parts contain Kaempferol-3β-D Galactoside, Quercitrin, Rutin, Quercetin, Ursolic acid, N-hexacosanol, Triacontane, Hyperoside, Pyrrolizidine, Alkaloids as Senkirkine, Cloronine, β-Sitosterol, Ralmitic acid, Stigmasterol.

Medicinal Properties:

शशश्रुतिः ज्वरहरा वातवैगुण्यनाशिनी ।। (हृदयप्रिया)

Śaśasṛutiḥ jvaraharā vātavaīguṇyanāśinī. (Hṛdayapṛiyā)

The plant is Antipyretic, Astringent, Opthalmic, Sudorific, Sweet, Thermogenic, and Vulnerary.

Medicinal Uses: The plant is useful in Asthma, Cuts, Diarrohea, Intermittent Fever, Ophthalmia, Otalgia, wounds.

Ethno- Medicinal Uses: Leaf – Boils (Kumar *et al.*, 1980); Eye diseases (Das *et al.*, 1983); Night blindness (Das *et al.*, 1983; Thomas & Britto, 2003); Wounds (Sharma, *et al.*, 1984; Tarafder, 1984); Cuts & Wounds (Shivakumar *et al.*, 2003); Sore ears (Thomas & Britto, 2003); Inflammation (Kumar *et al.*, 1980; Das *et al.*, 1983).

Root - Diarrhoea (Das *et al.*, 1983; Sinha *et al.*, 2006); Eye diseases, Inflammation (Pal & Srivastava, 1976); Wounds (Sinha *et al.*, 2006).

Whole plant - Eye complaint, Burns (Mukhopadhyay, 1992); Stiffneck (Udayan *et al.*, 2006); Diarrhoea, Wounds, (Mukhopadhyay, 1992; Sinha *et al.*, 2006).

EUPHORBIA THYMIFOLIA L. (EUPHORBIACEAE)

Regional & Other Names : *Bengali* – Didiya; *Gujarati* - Nahanidudheli; *Hindi* – Dudhi; *Kannada* – Kempu nene hakki; *Malayalam* - Nilppala; *Marathi* – Choti-dudhi; *Sanskrit* – Dugdhika; *Tamil* – Chinamampatchaiarisi; *Telugu* – Bidurumanasiyyam.

Distribution: Throughout India.

Taxonomic Description: An annual herb. Leaves very small and opposite. Involucers axillary, solitary or 2-3 in an axil. Capsules erect, obtusely keeled; styles short, 2-fid.

Medicinal Parts: Whole plant.

Chemical Constituents: Plant contains Epitaraxerol, N-hexacosanol, Euphorbol, Esters, N-alkanes, Sterols, Corilagin, Geraniin, Bixanin, Casuarin, Desgalloylstachyurin.

Medicinal Properties:

दुग्धिकोष्णा गुरु रूक्षा वातला गर्भकारिणी । (भा.प्र.)

Dugdhikoṣṇā guru rukshā vāttalā gạrbhkāriṇī. (Bhā. Pṛa.)

Acrid, Alexipharmic, Anthelmintic, Antibacterial, Antidiarrhoeal, Aphrodisiac, Astringent, Bitter, Demulcent, Depurative, Diuretic,

Expectorant, Laxative, Purgative, Vulnerary.

Medicinal Uses: Amenorrhoea, Asthma, Bronchitis, Bleeding, Piles, Cardiac debility, Chronic Cough, Dandruff, Dysentery, Dysmenorrhoea, Flatulence, Gonorrhoea, Helminthiasis, Leprosy, Skin diseases, Strangury, Wounds.

Ethno- Medicinal Uses: Whole plant - Blood dysentery (Jain,1984; Sinha *et al.*, 2006); Blood purifier (Mishra & Sahu, 1984); Ringworm, Skin diseases, Snake bite (Mishra & Sahu, 1984); Anthelminitc. (Shivkumar *et al.*, 2003); Leucorrhoea (Bhogaonkar & Kadam, 2006); Skin diseases (Sinha *et al.*, 2006); Diarrhoea & Dysentery (Kataria & Gauchan, 2006); Spermatrrhoea (Maheshwari *et al.*, 1980, 1981; Singh 2003; Prakash & Singh, 2006; Sinha *et al.*, 2006);

Latex - Acnes (Shivkumar *et al.*, 2003); Warts (Tigga *et al.*, 2003).

FERULA NARTHEX Boiss. (APIACEAE)

Regional & Other Names : *English* – Tibetian asafoetida; *Gujarati* - Hing; *Hindi* –Hing; *Sanskrit* – Sahastravedhi; *Tamil* – Perungayam.

Distribution: Baltistan & Astor.

Taxonomic Description: A perennial odorous herb.

Medicinal Parts: Niryaasa.

Chemical Constituents: Essential oil - Umbelliferon.

Medicinal Properties:

सहस्त्रवेधि जतुकं दाहिकं हिंगुरामठम् ।
हिगूष्णं पाचनं रूच्यं तीक्ष्णं वातबलासहृत् ।। (भा. प्र.)

Sahaṣravedhi jatukaṃ dāhikaṃ hiṇgurāmathaṃ
Hingūṣṇaṃ pācanaṃ rūcyaṃ tīkṣṇaṃ vātabalāsahṛt. (Bhā. Pṛa.)

Antiseptic, Antispasmodic, Emmenagogue, Expectorant, Nervine Stimulant.

Medicinal Uses: Asthma, Bronchitis, Flatulent colic, Pneumonia, Whooping cough.

Ethno- Medicinal Uses: Resin - Abortifacient (Singh *et al.,* 2003; Tiwari *et al.,* 1982).

FICUS ARNOTTIANA (Miq.) Miq. (MORACEAE)

Regional & Other Names : *Hindi* – Paras pipal; *Malayalam* - Kattarayal; *Sanskrit* – Plaksha; *Tamil* – Aswathom; *Telugu* – Kallaravi.

Distribution: Throughout India.

Taxonomic Description: A glabrous shrub or tree; Leaves broadly ovate, base cordate, Margin entire; Stipules ovate-lanceolate; Receptacles globose in pairs or clusters from tubercles in the axils of fallen leaves. Male flowers few, Gall & Fertile flowers indistinguishable.

Medicinal Parts: Bark.

Medicinal Properties:

परिशोऽन्यः पलाशश्च फलीशश्च कमण्डलुः ।
गर्दभाण्डः कन्दरालकपीतनसुपार्श्वकाः ।। (भा. प्र.)

Pariṣoaṇyaḥ palāṣaṣca phalīśaṣca kamaṇdaluḥ
Gaṛdabhāṇdaḥ kaṇdarālakapītanasupāṛṣvakāḥ. (Bhā. Pṛa.)

Bark - Aphrodisiac, Astringent, Constipating, Demulcent, Depurative, Emollient, Refrigerant, Sour, Sweet.

Medicinal Uses: Bark - Burning sensation, Diabetes, Diarrhoea. Inflammation, Leprosy, Pruritus, Scabies, Skin diseases, Wounds.

Ethno- Medicinal Uses: Leaf - Sterilizer (Lal & Lata, 1980).

FLACOURTIA JANGOMAS Rausch. (FLACOURTIACEAE)

Syn.: *F. cataphracta* Roxb.

Regional & Other Names : *Bengali* – Paniyala; *English* – Many spiked flacorita; *Gujarati* - Talispatra; *Hindi* – Talispatri; *Kannada* – Chankali; *Malayalam* - Talispatram; *Marathi* -Tambat; *Sanskrit* – Prachinamalaka; *Tamil* – Talispatri; *Telugu* – Talispatram.

Distribution: Throughout India.

Taxonomic Description: Large deciduous shrub or trees, with simple, ovate, glabrous leaves and Racemose Inflorescence.

Medicinal Parts: Bark, Leaves, Fruits.

Chemical Constituents: Ripe fruits - Reducing Sugars, Ash, Fat, Potassium, Tartaric acid, Proteins, Phenolics, Nitrogen, Iron, Phosphorus, Ascrobic acid and Unidentified chemicals. Gas chromatographic analysis of Fatty acids in Fats revealed the presence of palmitic, hexadecadienoic, Stearic, Oleic, Linoleic, □-linoleic and a few minor unidentifed acids. Amino acids from extract of their-dried ripe fruits showed the presence of Proline Hydroxyproline, Methtonine, Alanine, Glycine, and Valine. Paper Chromatography studies on simple reducing Sugars and their alditol acetates indicated the presence of Arabinose, Glucose, Fructose and Galactose.

Medicinal Properties:

गोपकण्टो रसे तिक्तः शीतलः शोफनाशनः ।
हन्ति श्लेष्माणमत्युग्रमुद्रक्तं हन्ति योगतः ।। (ध. नि.)

Gopakaṇto rase tiktah ̣sītalaḥ ṣophanāṣanaḥ
Haṇti ̣sleṣmāṇamaṭyugṛamudṛaktaṃ haṇti yogataḥ. (Dha. Ni.)

Bark, Leaves - Astringent, Acrid, Sour, Refrigerant, Stomachic, Diaphoretic.

Fruits - Sweet, Sour, Astringent, Acrid, Refrigerant, Digestive, Stomachic, Alexipharmic, Anti-Inflammatory, Depurative, Liver tonic.

Medicinal Uses: Bark, Leaves – Debility, Diarrhoea, Haemorrhoids, Stomatitis.

Fruits - Colic, Diabetes, Dyspepsia, Flatulence, Inflammation, Jaundice, Nausea, Rheumatism, Skin diseases.

Ethno- Medicinal Uses: Bark - Pre and postnatal for blood purifier (Kumar *et al.,* 1980).

GARUGA PINNATA Roxb. (BURSERACEAE)

Regional & Other Names : *Bengali* –Jum; *English* – Garuga; *Gujarati* - Khusimb; *Hindi* – Khogar; *Kannada* – Haiabalage; *Malayalam* - Annakkara; *Marathi* –Kakad; *Sanskrit* – Krsnamlika; *Tamil* – Yannaikkarai; *Telugu* – Gurugu.

Distribution: Throughout India.

Taxonomic Description: Large deciduous trees, with creamy White flowers and fleshy, glabrous, subpyriform drupes.

Medicinal Parts : Root, Stem, Leaves, Fruits, Galls.

Chemical Constituents: Leaves contain - Biphenyl ether-garuganin - I, alcohol - 6-Propyltetrdecan-7-ol. Stem bark contains - Resin contain - α - and β - amyrins, butyro-spermol, dammardienol.

Medicinal Properties:

गोलिकास्य फलं मधुरं शीतं दीपनपाचनम् ।। (स्व.)

"Golikāṣya phalaṃ madhuraṃ śītaṃ dīpanapācanaṃ." (Ṣva.)

Root - Pulmonary affections.

Fruits - Carminative, Cooling, Digestive, Sweet, Vermifuge.

Medicinal Uses: Leaf - Asthma.

Fruits – Gastropathy, Roundworm.

Galls - Foul ulcers, Obesity, Splenomegaly.

Ethno- Medicinal Uses: Leaf - Asthma (Shah *et al.,* 1981).

Bark – Diabetes, Wounds (Saxena *et al.,* 1981); Eye diseases (Sebastian & Bhandari, 1984); Piles, Skin infections (Gopal & Chandra, 2003); Eye diseases (Kapur *et al.*, 1992); Diabetes, Dysentery, Skin diseases (Sinha *et al.*, 2006).

Root - Skin diseases (Megoneitso, Rao, 1983; Sinha *et al.,* 2006); Diabetes, Dysentery (Sinha *et al.,* 2006).

Fruits - Dysentery (Roy & Chaturvedi, 1986; Sinha *et al.,* 2006); Diabetes, Skin diseases (Sinha *et al.,* 2006).

GENTIANA KURROO Royle. (GENTIANACEAE)

Regional & Other Names : *Bengali* – Karu; *English* – Indian gentian; *Gujarati* – Kadu chirayata; *Hindi* – Trayaman; *Kannada* – Karadihanni; *Malayalam* - Trayamana; *Sanskrit* – Trayamana; *Tamil* – Kampantirai; *Telugu* – Buroni.

Distribution: In Kashmir & the Himalayas, from 1,500 to 3,300 m.

Taxonomic Description: Perennial decumbent herb, with Blue spotted with White flowers and Capsules fruits.

Medicinal Parts: Rhizome.

Chemical Constituents: The drug sample from Kashmir on analysis yields 20% aqueous extract and 0.70% ash. Besides, a transparent, Brittle colourless and tasteless, resin has also been isolated in some extract from the plant.

Medicinal Properties:

त्रायन्ती शीतमधुरा गुल्मज्वरकफास्रनुत् ।
भ्रमतृष्णाक्षयग्लानिविषच्छर्दिविनाशिनी ।। (रा. नि.)

Ṭrāyaṇtī śītamadhurā guḷmajvarakaphāsṛanuṭ
Bhṛamaṭṛṣṇāksayaglānivisacchaṛdivināśiṇī. (Rā. Ni.)

Rhizome - Acrid, Anthelmintic, Anti-Inflammatory, Antileprotic, Bitter, Carminative, Depurative, Digestive, Diuretic, Febrifuge, Laxative, Sudorific, Thermogenic, Vulnerary.

Medicinal Uses: Rhizome - Amenorrhoea, Anorexia, Cardiac debility, Colic, Constipation, Dysmenorrhoea, Dyspepsia, Flatulence, Fever. Haemorrhoids, Helminthiasis, Inflammation, Leprosy, Leucoderma, Skin diseases, Strangury, Ulcers, Wounds.

Ethno- Medicinal Uses: Root - Tonic, Torpid liver (Nadkarni, 2002); Fever, Diabetic problems (Arya & Agarwal, 2006).

GREWIA TILIAEFOLIA Vahl. (TILIACEAE)

Regional & Other Names : *Bengali* – Dhamin; *English* – Dhaman; *Gujarati* - Dalmon; *Hindi* – Dhaman; *Kannada* – Dadasalu; *Malayalam* - Catacci; *Marathi* –Daman; *Sanskrit* – Dhanvangah; *Tamil* – Catacci; *Telugu* – Peddatena.

Distribution: Throughout India.

Taxonomic Description: Small trees, with ovate to orbicular leaves and yellow flowers in axillary cymes. Drupes globose.

Medicinal Parts: Bark, Fruits.

Chemical Constituents: Leaves contain - Tannin.

Medicinal Properties:

धन्वङ्गः कफपित्तास्रकासहृत्तुवरो लघुः ।
बृंहणो बलकृद्रूक्षः सन्धिकृद्व्रणेरोपणः ।। (भा. प्र.)

Dhanvaṇgaḥ kaphapiṭtāsṛakāsahṛṭtuvarō laghuḥ
Bṛṃhaṇo balakṛdṛūksaḥ saṇdhikṛdvṛaṇarōpaṇaḥ. (Bhā. Pṛa.)

Bark - Acrid, Antipruritic, Aphrodisiac, Astringent, Constipating, Emetic, Expectorant, Refrigerant, Styptic, Sweet, Tonic, Vulnerary.

Medicinal Uses- Bark - Burning sensation, Cough, Diarrhoea, General debility, Haemorrhages, Pruritus, Seminal weakness, Skin diseases, Ulcers, Wounds.

Ethno- Medicinal Uses: Leaves - Normal delivery (Ravikumar & Sankar, 2003).

Root - Syphilis (Chaudhuri *et al.,* 1982; Sinha *et al.,* 2006).

Bark - Dysentery (Basak, 2006).

HABENARIA EDGEWORTHII Hook. f. ex. Collett. (ORCHIDACEAE)

Regional & Other Names : *Hindi* Rddhi; *Kannada* – Rddhi; *Malayalam* - Rddhi; *Sanskrit* – Rddhi; *Tamil* – Rddhi; *Telugu* – Rddhi.

Distribution: In Shimla, Garhwal, U.P. at about 2,500 m.

Taxonomic Description: Tuberous terrestrial herbs, with sheating ovate-lanceolate leaves and Yellowish Green flowers.

Medicinal Parts : Tuber.

Medicinal Properties:

ऋद्धिर्बल्या त्रिदोषघ्नी शुक्लला मधुरा गुरूः ।
प्राणैश्वर्यकरी मूर्च्छारक्तपित्तविनाशिनी ।। (भा. प्र.)

Rddhiṛbaḷyā ṭridosaghnī śuklalā madhurā guruḥ
Pṛaṇaiṣvaṛyakarī mūṛccharaktapiṭtavināśinī. (Bhā. Pṛā.)

Tuber - Anthelmintic, Aphrodisiac, Appetiser, Depurative, Emollient, Refrigerant, Rejuvenating, Sweet.

Medicinal Uses: Tuber - Anorexia, Asthma, Burning sensation, Cough, Fever, General debility, Gout, Helminthiasis, Leprosy, Skin diseases.

HABENARIA INTERMEDIA D. Don. (ORCHIDACEAE)

Regional & Other Names : *Hindi* – Vrddhi; *Kannada* – Vrddhi ; *Malayalam* - Vrddhi; *Sanskrit* – Vrddhi; *Tamil* – Vrddhi; *Telugu* – Vrddhi .

Distribution: In the Himalayas.

Taxonomic Description: Terrestrial plants, with sheating ovate-oblong leaves and Greenish-White flowers.

Medicinal Parts: Tuber.

Medicinal Properties:

वृद्धिस्तु मधुरा वृष्या शीता गर्भबलप्रदा ।
बृहणी कफपित्तास्रक्षतकासक्षयापहा ।। (कै. नि.)

Vṛddhiṣtu madhurā vṛṣyā śītā gaṛbhabalapṛadā
Bṛmhaṇī kaphapiṭtāṣraksatakāsaksayāpahā. (Kaī. Ni.)

Tuber - Aphrodisiac, Depurative, Emollient, Anthelmintic, Appetiser. Refrigerant, Rejuvenating, Sweet.

Medicinal Uses: Tuber - Anorexia, Asthma, Burning sensation, Cough, Fever, General debility, Gout, Helminthiasis, Leprosy, Skin diseases.

HEDYOTIS HERBACEA L. (RUBIACEAE)

Syn.: *Oldenlandia herbacea* (L.) Roxb.

Regional & Other Names : *Hindi* – Paper-bhed; *Malayalam* - Nonnanampullu; *Sanskrit* – Chayaparpatika; *Tamil* – Nonnanampullu; *Telugu* – Chiviveru.

Distribution: Throughout India.

Taxonomic Description: An erect bushy annual or biennial; leaves simple, linear-lanceolate, opposite; flowers White, small and fruits capsules.

Medicinal Parts: Whole plant.

Chemical Constituents: Kaempferol-3-O-Rutinoside and ursolic acid were Isolated from the leaves & tiwgs.

Medicinal Properties:

छायापर्पटिका शीता सतिक्ता मधुरा लघुः ।
गुल्मशोफज्वरहरा कृमिवातविकारहा ।। (स्व.)

"Chāyāpaṛpatikā śītā satiktā madhurā laghuḥ
Guḷmaśophajvaraharā kṛmivātavikārahā." (Ṣva.)

Plant - Anthelmintic, Anti-Inflammatory, Bitter, Cooling, Expecotrant, Febrifuge, Stomachic, Sweet.

Medicinal Uses: Plant - Bronchitis, Elephantiasis, Fever, Hydrocele, Ulcers.

Ethno- Medicinal Uses: Whole plant - Body pain (Ramachandran & Nair, 1981).

HEDYCHIUM SPICATUM Buch.-Ham. ex Smith (ZINGIBERACEAE)

Regional & Other Names : *Bengali* – Grandhasthati; *English* – Spiked ginger lily; *Gujarati* - Kapurakachari; *Hindi* – Kapurkachri; *Kannada* – *Gandhashati*; *Marathi* –Kapurakachari; *Sanskrit* –Kapurakachali; *Tamil* – Simai-kkichiklikkishangu.

Distribution: China, Nepal, Kumaon, Subtropical Himalayas.

Taxonomic Description: A perennial rhizomatous herb or a tall tree with root stock.

Medicinal Parts: Rhizome, Root stock.

Chemical Constituents: Rhizome - Essential oil, Starch, Carbonic acid, a Glycoside.

Medicinal Properties:

शटी स्यात् कटुतीक्ष्णोष्णा सन्निपातज्वरापहा ।
कफास्त्रव्रणकासघ्नी वक्त्रशुद्धिविधायिनी ।। (ध. नि.)

Śatī ṣyāṭ katutīkṣṇoṣṇā saṇnipātajvarāpahā
Kaphāṣṛavṛaṇakāsaghnī vaktraṣuddhividhāyinī. (Dha. Ni.)

Root stock - Acrid, Astringent, Bitter, Laxative, Pungent.

Rhizome - Carminative, Stimulant, Stomachic, Tonic.

Medicinal Uses: Root stock - Asthma, Bronchitis, Diseases of blood, Hiccough, Inflammation, Liver complaints, Pains, Vomiting.

Rhizome - Dyspepsia.

Ethno- Medicinal Uses: Root stock - Snake bite, Scorpion sting (Rai *et al.*, 1992); Asthma, Internal injury (Arya & Prakash, 2003).

Rhizome - Asthma (Malhotra & Balodi, 1984; Sharma *et al.*, 1979); Bronchitis (Sharma *et al.*, 1979; Chhetri, 2005); Eye diseases (Megoneitso, Rao, 1983); Gastric diseases, Blood purifier (Malhotra & Balodi, 1984); Tonic (Singh *et al.*, 1992); Anti-inflammatory (Kapur *et al.*, 1992); Stomachic (Janmeda *et al.*, 2006); Tuberculosis (Chhetri, 2005).

HELIOTROPIUM INDICUM L. (BORAGINACEAE)

Regional & Other Names : *Bengali* – Hathisura; *English* – Indian turnsole; *Gujarati* – Hathi-sundhana; *Hindi* – Siriyari; *Kannada* – Celubaladagidha; *Malayalam* - Telkkata; *Marathi* –Bhurundi; *Sanskrit* – Vrscikali; *Tamil* – Telkodukkai; *Telugu* – Kodikki.

Distribution: Throughout India.

Taxonomic Description: Erect herbs, pale-violet flowers and fruit of 1-seeded nutlets.

Medicinal Parts: Whole plant.

Chemical Constituents: Aerial parts of the plant contains as Indicine, Echinatine, Supinine, Heleurine, Heliotrine, Lasiocarpine, Supinidine, Retranecine, Lindelofidine, Indicine-N-oxide, Indicinine, Supinene, Rapanone, Lupeol, N-hexacosanol, Sitosterol, Stigmasterol, Campesterol, Cholinasterol.

Medicinal Properties:

वृश्चिकाली कटुस्तिक्ता सोष्णां हृद्वक्त्रशुद्धिकृत् ।
रक्तपित्तहरा बल्या विबंधारोचकापहा ।। (रा. नि.)

Vṛścikāli katuṣtiktā soṣṇāṃ hṛdvakṭraṣuddhikṛt
Raktapiṭtahara baḷyā vibaṇdhārocakāpahā. (Rā. Ni.)

Plant - Astringent, Bitter, Diuretic, Emollient, Thermogenic, Vulnerary.

Root - Aphrodisiac, Astringent, Bitter, Expectorant, Febrifuge.

Medicinal Uses: Plant - Wounds, Skin affections, Rheumatism, Ulcers.

Root - Cough, Bornchitis, Fever, Ringworm, Rheumatism.

Leaves - Fever, Urticaria, Ulcers, Wounds, Gonorrhoea, Ringworm, Rheumatism, Erysipelas.

Ethno- Medicinal Uses: Whole plant –Insect stings (Shah *et al.,* 1983; Singh & Pandey, 1980); Throat infection, ulcer (Shah *et al.,* 1983); Wounds, Ulcers (Mukhopadhyay, 1992); Eye inflammation (Sinha *et al.,* 2006). Boils (Shah *et al.,* 1983; Mukhopadhyay, 1992; Singh & Pandey, 1980; Sinha *et al.,* 2006).

Flowers - Abortifacient (Tarafder, 1983); Boils, Eye inflammation (Sinha *et al.,* 2006).

Leaf - Boils (Dwivedi & Pandey, 1992); Insect stings (Kapur *et al.,* 1992).

Fruit - Boils, Wounds (Deokule, 2006).

Root - Ear troubles (Das *et al.,* 2006).

HELLEBORUS NIGER L. (RANUNCULACEAE)

Regional & Other Names : *Bengali* – Kalakutki; *English* – Black hellebore; *Gujarati* - Koddu; *Hindi* – Kalikatuki; *Sanskrit*-Katuki; *Tamil* – Katurohini; *Telugu* – Katukarohini.

Distribution: Occurring in Himalayas from Kashmir to Sikkim.

Taxonomic Description: A Rhizomatous herb.

Medicinal Parts: Rhizome, Root, Whole plant.

Chemical Constituents: Rhizome - Helleborin, Hellebrin, Helleborein, Saponin, a Volatile oil, Glycosides.

Medicinal Properties:

कटुका पित्तजित्तिक्ता कटुः शीतास्त्र दाहजित् ।
बलासारोचकान् हन्ति विषमज्वर नाशिनी ।। (ध. नि.)

Katukā pịttajịttiktā katuḥ śītāṣra dāhajịt
Balāsārocakāṇ haṇti visamajvara nāśinī. (Dha. Ni.)

Rhizome - Anthelmintic, Cathartic, Diuretic, Emmenagogue.

Medicinal Uses: Rhizome - Amenorrhoea, Cardiac tonic, Chlorosis, Dropsy, Hysteria, Nervous disorders, Skin diseases.

HIBISCUS ACULEATUS Roxb. (MALVACEAE)

Regional & Other Names : *Kannada* – Bendhegida; *Malayalam* - Paniccakam; *Sanskrit* – Sathambasthi; *Tamil* – Kontagomgura; *Telugu* – Bendakaya.

Distribution: Throughout India.

Chemical Constituents: Seed oil contain - Palmitic, stearic, oleic, linoleic, 12,13-epoxyolek acids. Flowers contain - A flavonolglucoside-hibiscatin.

Taxonomic Description: A trailing prickly herb, with large Purple Centred Yellow flowers and ovoid pointed capsules.

Medicinal Parts: Root, Leaves.

Medicinal Properties:

शठाम्बष्ठी पत्रं कषायाम्लं लघुर्नेत्रहितं हिमम् ।
पक्वातिसारकृच्छ्रघ्नं ज्वरकृमिविनाशनम् ।। (स्व.)

"Śathāmbaṣthī patṛaṃ kasāyaṃlaṃ laghurneṭrahitaṃ himaṃ
Pakvatisārakṛcchṛaghnaṃ jvarakṛmivināśanaṃ." (Ṣva.)

Root - Anti-Inflammatory, Diuretic, Refrigerant.

Leaf - Anthelmintic, Anti-Inflammatory, Astringent, Sour.

Medicinal Uses: Root - Burning sensation, Inflammation.

Leaf - Dyspepsia, Helminthiasis, Inflammation.

HIBISCUS CANNABINUS L. (MALVACEAE)

Regional & Other Names: *Bengali* – Machika; *English* – Mesta; *Gujarati* - Ambari; *Hindi* – Pitwa; *Kannada* – Nalita; *Malayalam* - Kanchava; *Marathi* – Ambada; *Sanskrit* – Anvastha; *Tamil* – Kasini; *Telugu* – Gaynaru.

Distribution: Native of Africa; cultivated for its silky bast fibres.

Taxonomic Description: Annual or perennial, prickly herbs. corolla spreading, yellow with crimson centre.

Medicinal Parts: Flower, Seed, Leaves.

Chemical Constituents: Seed - Phospholipids, Palmitic acid, Stearic acid, Oleic acid, Linolic acid, β-sitosterol, Limonene, Citral.

Leaves - Aldobiouronic acid, Rutin, Isoquercitrin, Limonene, Phellandrene, □-terpenyl acetate, Lignoceric acid, Cerotic acid, Montanic acid, Melissic acid.

Flower - Anthocyanin, Fructose, Glucose, Mannose, Sucrose, Sterculic acid, Tridecanoic acid.

Medicinal Properties:

माचिकाऽमला रसे पाके कषाया शीतला लघुः ।
पक्कातीसारपित्तास्रकफकण्ठामयापहा ।। (भा. प्र.)

Māchikāamlā rase pāke kaṣāyā śitalā laghuḥ
Pakkātisārpittāsrakaphakanthāmayapahā (Bha. Pra.)

Seeds - Analgesic, Aphrodisiac, Appetising, Stomachic.

Leaves - Purgative.

Flower - Antibilious, Antipyretic, Antiulcer, Digestive.

Medicinal Uses: Flower - Abdominal pain, Biliousness, Digestion, Dysentery, Fever.

Ethno- Medicinal Uses: Leaf - Purgative (Janardhanan, 1963).

HOLOSTEMMA ADA-KODIEN Schultes. (ASCLEPIADACEAE)

Syn.: *H. annulare* (Roxb.) K. Schumn.; *H. rheedei* Wall.

Regional & Other Names : *Gujarati* - Kharner; *Hindi* – Chirvel; *Malayalam* - Atapatiyan; *Marathi* – Dudurli; *Sanskrit* – Jivanti; *Tamil* – Palaikkirai; *Telugu* – Palagurugu.

Distribution: Throughout India.

Taxonomic Description: Laticiferous twining shrubs, with Purple flowers and foliicles fruits.

Medicinal Parts: Roots.

Chemical Constituents: Root tubers contain □-amyrin, Lupeol, β-sitosterol, Alanine, Aspartic acid, Glycine, Serine, Thereonine, Valine, Calcium, Phosphate.

Medicinal Properties:

जीवन्ती शीतला स्वादुः स्निग्धा दोषत्रयापहा ।
रसायनी बलकरी चक्षुष्या ग्राहिणी लघुः ।। (भा. प्र.)

Jīvaṇtī śītalā ṣvāduḥ ṣnigdhā dosaṭrayāpahā
Rasāyanī balakarī caksuṣyā gṛāhiṇī laghuḥ. (Bhā. Pṛa.)

Roots - Alterant, Aphrodisiac, Emollient, Expectorant, Ophthalmic, Refrigerant, Stimulant, Sweet, Tonic.

Medicinal Uses: Roots - Burning sensation, Consumption, Cough, Fever, Ophthalmopathy.

Ethno- Medicinal Uses: Latex - Ulcers (Shah *et al.*, 1983; Sinha *et al.*, 2006); Labour pain, Veneral diseases (Sinha *et al.*, 2006).

Root - For lactation (Saxena, 1986); Ulcer, Labour pain, Veneral diseases (Sinha *et al.*, 2006).

Flower - Veneral diseases (Jain & De, 1966; Sinha *et al.*, 2006); Ulcer, Labour pain, (Sinha *et al.*, 2006).

HOMONOIA RIPARIA Lour. (EUPHORBIACEAE)

Regional & Other Names : *Bengali* – Jalavetasah; *Hindi* – Jalbent; *Kannada* – Sannapasanabheda; *Malayalam* - Katallari; *Sanskrit* – Jalavetasah; *Tamil* – Kattuarali; *Telugu* – Siridamanu.

Distribution: Throughout India.

Taxonomic Description: Deciduous shrubs, with spikes inflorescens and capsule fruits.

Medicinal Parts: Roots, Leaves, Fruits.

Chemical Constituents: Root contains - □-spinasteryl acetate, Fatty acids like myristic acid, Palmitic acid, Stearic acid, Oleic acid.

Leaves contain Quercetain-3-O-β-D-Glucopyranosyl (1→6) -□-L-Rhamnoside, Gallic acid, Taraxerone.

Medicinal Properties:

जलजो वेतसः शीतो संग्राही वातकोपनः ।। (भा. प्र.)

Jalajo vetasaḥ śītlo saṃgrāhī vātakopanaḥ. (Bhā. Pṛa.)

Roots - Depurative, Diuretic, Emetic, Laxative.

Refrigerant, Leaves / Fruits – Antiseptic, Depurative.

Medicinal Uses: Roots - Haemorrhoids, Vesical Calculi, Strangury, Gonorrhoea, Urethrorrhea, Syphilis.

Leaves / Fruits - Ulcers, Wounds, Skin diseases.

Ethno- Medicinal Uses: Root - Back pain (Binu *et al.,* 2003); Malaria fever, Diuretic (Sinha *et al.,* 2006); Urinary infection (Nadanakunjidam, 2003).

HUGONIA MYSTAX L. (LINACEAE)

Regional & Other Names : *English* – Climbing flax; *Kannada* – Mrema; *Malayalam* - Karttotti; *Sanskrit* – Kamsamarah; *Tamil* – Motirakkanni; *Telugu* – Kakibira.

Distribution: Throughout India.

Taxonomic Description: Shrub with Yellow tomentose twigs; Leaves simple elliptic-obovate glabrous; flowers Yellow; fruits drupes.

Medicinal Parts: Roots.

Medicinal Properties:

कंसमारः कषायतिक्तः सुस्वादुः शोफनाशनः ।। (स्व.)

"Kaṇsamāraḥ kasāyatiktaḥ suṣvāduḥ śophanāṣanaḥ." (Ṣva.)

Roots – Anthelmintic, Astringent, Bitter, Febrifuge, Sweet.

Medicinal Uses: Roots - Fevers, Inflammations, Verminosis.

Ethno- Medicinal Uses: Leaf - Internal swellings of wounds (Prasad *et. al.,* 2003).

HYGRORYZA ARISTATA (Retz.) Nees ex Wight & Arn. (POACEAE)

Syn.: *Pharus aristatus* Retz.

Regional & Other Names : *Bengali* – Uridhan; *English* – Bengal wild rice; *Hindi* – Jangali-Dal; *Kannada* –Jyarahumedhe; *Malayalam* - Varinellu; *Marathi* – Deobhata; *Sanskrit* – Nivarah; *Tamil* – Vallippul .

Distribution: Throughout India.

Taxonomic Description: A glabrous floating grass. Stem spongy with feathery whorted roots at the nodes. Leaves linear or ovate-oblong, with inflated sheaths. Spikelets narrow; 1-flowered sessile or pedicellate. Floral glume lanceolate awned; palea as long as the glume, awnless, containing a bisexual floret. Lodicules 2, stamen 6, style 2, free. Grain narrowly oblong.

Medicinal Parts: Roots, Seeds.

Chemical Constituents: Analysis of grass for free amino acids and amino acids in peptides and protein hydrolysates; protein content of grass 7.92%.

Medicinal Properties:

नीवारः शीतलो ग्राही पित्तघ्नः कफवात्कृत् ।। (भा. प्र.)

Nīvāraḥ ṣītalo grāhī piṭtaghnaḥ kaphavāṭakṛṭ. (Bhā. Pṛa.)

Root / Seed - Acrid, Cooling, Diuretic, Emollient, Constipaitng, Sweet, Tonic.

Medicinal Uses: Root / Seed - Burning sensation, Diarrhoea, Fatigue, General debility, Otopathy, Strangury.

ILLICIUM VERUM Hook. f. (MAGNOLIACEAE)

Regional & Other Names : *English* – Star anise of China; *Hindi* – Anasphal; *Kannada* – Kankola; *Malayalam* - Takkolapputtil; *Sanskrit* – Takkolakam; *Tamil* – Anusappu; *Telugu* – Kuppi.

Distribution: Indigenous to China.

Taxonomic Description: Evergreen trees, with White or Red flowers and Star shaped fruits.

Medicinal Parts: Fruits.

Chemical Constituents: Fruit contains a Volatile essential oil, Sugar, a bitter Principle and tannin.

Medicinal Properties:

तक्कोलं दीपनं पाचनं स्वादु रोचनं कफनाशनम् ।। (स्व.)

"Takkolaṃ dīpanaṃ pācanaṃ śvadu rōcanaṃ kaphanāśanaṃ." (Ṣva.)

Fruits - Aromatic, Carminative, Digestive, Expectorant, Stomachic, Sweet.

Medicinal Uses: Fruits - Asthma, Cough, Dysentery, Dyspepsia, Facial paralysis, Flatulence, Spasmodic.

IPOMOEA PES-CAPRAE (L.) R. Br. (CONVOLVULACEAE)

Syn.: *I. biloba* Forssk.; *I. maritima* R. Br.

Regional & Other Names : *Bengali* – Chhagalkuri; *English* – Goat's foot creeper; *Gujarati* - Marjadavela; *Hindi* – Dopatilata; *Kannada* – Bangadivalli; *Malayalam* - Atampuvalli; *Marathi* – Maryadvel; *Sanskrit* – Maryadavalli; *Tamil* – Atampu; *Telugu* – Balabantatige.

Distribution: Throughout India.

Taxonomic Description: Perennial herbs, with Purpled-Red flowers and capsules fruits.

Medicinal Parts: Whole plant.

Chemical Constituents: Plant contains Pentatriacontane, Triacontane Behenic acid, Melissic acid, Butyric acid, Myristic acid, Malic acid, Citric acid, Tartaric acid, Succinic acid, Fumaric acid, Volatile oil, Indole alkaloids, Quercetin-3-glucoside, Quercetin-3-Galactoside.

Medicinal Properties:

मर्यादवल्लिका शीता ग्राहिणी सारका गुरूः ।
पाककाले चोष्णा स्याद्वातला गर्भकर्षिनी ।।
विषूचिकां च शूलं च वान्ति चामं च नाशयेत् ।। (नि. र.)

Maryādavaḻlikā śītā g̣rāhiṇī sārakā gurūḥ
Pākakāle coṣaṇā ṣyādvātalā gaṛbhakaṛsinī
Visūcikāṃ ca śūlaṃ ca vāṇti cāmaṃ ca nāśayeṭ. (Ni. Ra.)

Whole plant - Acrid, Astringent, Diuretic, Laxative, Refrigerant, Stomachic, Tonic.

Medicinal Uses: Whole plant - Boils, Burning sensation, Carbuncle, Colic, Dropsy, Dyspepsia, Flatulence, Haemorrhoids, Menorrhagia, Skin diseases, Strangury, Swellings, Ulcers, Vomiting, Wounds.

Ethno- Medicinal Uses: Leaf - Rheumatism (Shah *et al.,* 1983).

Juice- Purgative, Skin infections (Tigga *et al.,* 2003).

IPOMOEA SEPIARIA Roxb. (CONVOLVULACEAE)

Syn.: *I. maxima* auct. non (L. f.) Sweet.

Regional & Other Names : *Bengali* –Bankalmi; *Gujarati* -Hanumanvel; *Hindi* – Bankalmi; *Kannada* – Laksmana; *Malayalam* - Tirutali; *Marathi* – Amtivel; *Sanskrit* – Laksmana; *Tamil* – Talikkirai; *Telugu* – Laksmana.

Distribution: Throughout India.

Taxonomic Description: Perennial twining plants, with Pink flowers and capsule fruits.

Medicinal Parts: Whole plant.

Medicinal Properties:

लक्ष्मणा गर्भदा शीता सरा वृष्या त्रिदोषनुत् ।। (म. पा. नि.)

Lakṣmaṇā gaṛbhadā śītā sarā vṛṣyā ṭridosanuṭ. (Ma. Pā. ni.)

Whole plant - Aphrodisiac, Cooling, Diuretic, Laxative, Rejuvenating, Sweet, Tonic.

Medicinal Uses: Whole plant - Burning sensation, General debility, Strangury.

Ethno- Medicinal Uses: Root - Diabetes (Saxena *et al.,* 1981).

IXORA COCCINEA L. (RUBIACEAE)

Regional & Other Names : *Bengali* – Rangan; *English* – Jungle-flame ixora; *Hindi* – Rangan; *Kannada* – Kiskara; *Malayalam* - Tecci; *Marathi* – Bakora; *Sanskrit* – Paranti; *Tamil* – Cetti; *Telugu* – Manmadibanum.

Distribution: South-West Peninsular India.

Taxonomic Description: A shrub. Leaves oval with cordate, amplexicaul base. Flowers deep Red, attractive, in Corymbs. Berries hard.

Medicinal Parts: Root, Leaves, Flowers.

Chemical Constituents: Root - Mannitol, Myristic acid, Cyanidin, Delphinidin, Methyl esters of Palmitic acid, Stearic acid, Oleic acid, Linoleic acid.

Flowers - Rutin, Cyanidin-3-rutinoside, Leucocynidine Glycoside.

Medicinal Properties:

किंशुकः कषायतिक्तः मधुरः ग्राही दीपनपाचनः ।। (स्व.)

"Kiṃṣukaḥ kasāyatiktaḥ madhuraḥ gṛāhī dīpanapācanaṃ." (Ṣva.)

Root - Acrid, Antiseptic, Astingent, Febrifuge, Sedative, Stomachic.

Flowers - Astringent, Bitter, Carminative, Constipating, Digestive, Sweet.

Medicinal Uses: Root - Chronic ulcers, Diarrhoea, Dysentery, Fever, Gonorrhoea, Hiccough, Skin diseases, Sores.

Leaves - Diarrhoea.

Flowers - Dysentery, Dysmenorrhoea, Leucorrhoea, Sores, Ulcers.

Ethno- Medicinal Uses: Root - Dysentery (Kamble & Pradhan, 1980).

Flowers - Dysentery (Deokule, 2006).

JASMINUM ANGUSTIFOLIUM (L.) Willd (OLEACEAE)

Regional & Other Names : *English* – Wild Jasmine; *Hindi* – Banmallika; *Kannada* – Adavimallige; *Malayalam* - Kattumulla; *Sanskrit* – Asphota; *Tamil* – Kattumalligai; *Telugu* – Adavimali.

Distribution: Circars, Deccan, Carnatic, on Eastern side, down to Travancore.

Taxonomic Description: Small scandent Shrubs, with Simple, ovate leaves and Berries fruits.

Medicinal Parts: Root, Leaves.

Chemical Constituents: Flowers - Indole.

Medicinal Properties:

वनमल्लिका कटुतिक्ता च चक्षुष्या मुखपाकहृत् ।। (स्व.)

"Vanamaḷlikā katutiktā ca caksuṣyā mukhapākahṛt." (Ṣva.)

Root – Acrid, Bitter.

Leaves - Eemtic.

Medicinal Uses: Root - Leprosy, Pruritus, Ulcerative Stomatitis, Wounds.

Ethno- Medicinal Uses: Root - Antidote for poison (Kumar & Pullaiah, 2003); Ringworm (Subramaniam, 2003).

JASMINUM ARBORESCENS Roxb. (OLEACEAE)

Syn.: *J. roxburghianum* Wall.

Regional & Other Names :*Bengali* – Chameli; *English* – Tree jasmine; *Gujarati* - Chameli; *Hindi* – Bela, chameli; *Kannada* – Visamallige; *Malayalam* - Nagamallika; *Sanskrit* – Vasanti, Nagamalli; *Tamil* – Nagamalli; *Telugu* – Adavimalli.

Distribution: Throughout India.

Taxonomic Description: A large shrub with hairy climbing branchlets; leaves simple, opposite, ovate-oblong; flowers White, Fragrant, panicled; fruits Berry.

Medicinal Parts: Leaves.

Chemical Constituents: Oil contains Benzyl benzoate, Benzyl alcohol, Eugenol, Farnesot, Bergamotene, Nerolidol, P-cresol, Benzyl acetate, Bergamontane, Geranoil, Indole, Linalool, Benzoic acid, Vanilioin.

Medicinal Properties:

वासन्ती शीतला तिक्ता लध्वी दोषपत्रयापहा ।
अत्यन्तशीतलं कन्दमीषत् कटुरसं स्मृतम् ।। (कै. नि.)

Vāsanti śītalā tiktā laghvi doṣapatrayāpahā
Atyaṇtśītalaṃ kaṇdmīsaṭ katurasaṃ sṃritaṃ (Kai. Ni.)

Leaves - Astringent, Bitter, Emetic, Refrigerant, Stomachic, Sweet, Tonic.

Medicinal Uses: Leaves - Colic, Cough, Faltulence, General debility, Menstrual disorders.

Ethno- Medicinal Uses: Root - Fever (Tarafder, 1986; Sinha *et al.,* 2006); Asthma, Bronchitis (Sinha *et al.,* 2006).

JASMINUM AURICULATUM Vahl. (OLEACEAE)

Regional & Other Names : *Bengali* – Umbustha; *English* – Needle flower Jasmine; *Hindi* – Juhi; *Kannada* – Kadaramallige; *Malayalam* - Tusimulla; *Sanskrit* – Yuthika; *Tamil* – Usimalligai; *Telugu* – Adavimulla.

Distribution: Throughout South India.

Taxonomic Description: A bushy or climbing shrub. Leaves with a pair of minute leaflets or simple. Flowers White, Star shaped.

Medicinal Parts: Root, Flowers.

Chemical Constituents: Indole, Benzyl benzoate, Methyl Anthranilate, Benzyl Alcohol.

Medicinal Properties:

यूथिकायुगलं स्वादु शिशिरं शर्करार्त्तिनुत् ।
पित्तदाहतृषाहारि नानात्वग्दोषनाशनम् ।। (रा. नि.)

Yūthikāyugalaṃ ṣvādu ṣiṣiraṃ ṣaṛkarāṛttinuṭ
Pịttadāhatṛsāhārinānāṭvagdosanāṣanaṃ. (Rā. Ni.)

Flowers - Acrid, Astringent, Bitter, Depurative, Diuretic, Fragrant, Refrigerant, Sweet.

Medicinal Uses: Root - Skin disease.

Flowers - Burning sensation, Hyperdipsia, Ophthalmopathy, Strangury, Ulcers.

Ethno- Medicinal Uses: Leaf - Tonic (Sinha *et al.*, 2006).

JASMINUM MULTIFLORUM (Burm. f.) Andr. (OLEACEAE)

Syn.: *J. pubescens* Willd.; *J. hirsutum* Willd.

Regional & Other Names : *Bengali* – Kund; *English* – Downy jasmine; *Gujarati* - Dolor; *Hindi* – Kundphul; *Kannada* – Magimallige; *Malayalam* - Kurukkuttimulla; *Marathi* – Mogra; *Sanskrit* – Kundah; *Tamil* – Malligai; *Telugu* – Gajari.

Distribution: Thorughout India.

Taxonomic Description: A scandent or diffuse, densely pubescent shrub. Leaves simple, ovate thick, cordate. Flowers White slightly scented in terminal, many-flowered umbles. Calyx teeth 1 cm. or more in length.

Medicinal Parts: Leaves, Flowers.

Chemical Constituents: Leaves - Friedelin, Lupeol, Betulin, □-amyrin, Betulinic acid, Ursolic acid, Oelanolic acid, Jasmolactones, A,B,C & D, Multiforoside, Multiroside, Jasmutiside, Multiflorin.

Medicinal Properties:

कुन्दोऽतिमधुरः शीतः कषायः कैश्यभावनः ।
कफपित्तहरश्चैव सरो दीपनपाचनः ।। (रा. नि.)

Kuṇdoatimadhuraḥ śītaḥ kasāyaḥ kaīśyabhāvanaḥ
Kaphapiṭtaharaṣcaīva saro dīpanapācanaḥ. (Rā. Ni.)

Flowers - Acrid, Alexipharmic, Bitter, Depurative, Digestive.

Laxative, Refrigerant, Sweet.

Medicinal Uses: Flowers – Cephalagia, Inflammation, Rheumatism.

Ethno- Medicinal Uses: Bark - Burns (Sinha *et al.*, 2006; Maheshwari *et al.*, 1980, 1981).

Leaf - Fever, Typhoid (Pereira, 2006); Headache (Pereira, 2006).

JUSTICIA BEDDOMEI (Clarke.) Bennet (ACANTHACEAE)

Syn.: *Adhatoda beddomei* C.B. Clarke

Regional & Other Names : *Hindi* – Adusa; *Kannada* – Sanna adusoge; *Malayalam* - Atalotakam; *Sanskrit* – Vasa; *Tamil* – Adutota; *Telugu* – Addasaramu.

Distribution: In Kerela, in the Hilly regions.

Taxonomic Description: Large shrubs, with condensed spikes and capsule fruits.

Medicinal Parts: Whole plant.

Medicinal Properties:

> वाशा तिक्ता कटुः शीता कासघ्नी रक्तपित्तजित् ।
> कामलाकफवैकल्यज्वरश्वासक्ष्यापहा ।। (रा. नि.)
>
> Vāṣā tiktā katuḥ śītā kāsaghnī raktapiṭtajiṭ
> Kāmalākaphavaīkaḷyajvaraṣvāsakaṣayāpahā. (Rā. Ni.)

Whole plant - Antispasomdic, Astringent, Bitter, Depurative, Diuretic, Expectorant, Febrifuge, Refrigerant, Styptic.

Medicinal Uses: Leaves - Irritable cough, for bleeding in diarroea, Haemoptysis.

Flower - Ophthalmia.

Ethno- Medicinal Uses: Leaves - Cough, Cold (Arinathan *et al.,* 2006).

KAEMPFERIA GALANGA L. (ZINGIBERACEAE)

Regional & Other Names : *Bengali* – Chandumula; *Gujarati* - Kapurkachri; *Hindi* – Kandramula; *Kannada* – Kacora; *Malayalam* - Kaccuri; *Marathi* – Kachri; *Sanskrit* – Sathi; *Tamil* – Kaccolam; *Telugu* – Chandramula.

Distribution: Cultivated throughout India.

Taxonomic Description: Aromatic perennial herbs, with Purplish dotted white flowers and Capsule fruits.

Medicinal Parts: Rhizome, Root Stock, Leaves.

Chemical Constituents: Tuber contains an alkaloid, Starch, Gum, Fatty matter with a fragrant liquid Essential oil and a Solid White Crystalline substance & Mineral matter.

Medicinal Properties:

कर्चूरः कटुतिक्तोष्णः कफकासविनाशनः ।
मुखवैशद्यजननो गलगण्डादिदोषनुत् ।। (रा. नि.)

Karcūraḥ katutiktoṣṇaḥ kaphakāsavināśanaḥ
Mukhavaīṣadyajanano galagaṇdādidosanuṭ. (Rā. Ni.)

Rhizome / Root stocks- Acird, Anthlemintic, Aromatic, Bitter, Carminative, Depurative, Digestive, Diuretic, Expectorant, Febrifuge, Stimulant, Thermogenic, Vulnerary.

Medicinal Uses: Rhizome / Root stocks - Asthma, Bronchitis, Cough, Dyspepsia, Fever, Leprosy, Rheumatism, Skin diseases, Strangury, Ulcers, Wounds.

Leaves - Fever, Ophthalmia, Rheumatism, Swellings.

Ethno- Medicinal Uses: Rhizome - Tonsillitis (Singh *et al.,* 1992).

LAGERSTROEMIA SPECIOSA (L.) Pers. (LYTHRACEAE)

Syn.: *L. flosreginae* Retz.

Regional & Other Names : *Bengali* – Jarool; *English* – Pride of India; *Hindi* – Jarul; *Kannada* –Hole-dasavala; *Malayalam* – Manimaruthu; *Marathi* – Taman; *Sanskrit* – Tinisa; *Tamil* –Kadali ; *Telugu* –Varagogu.

Distribution: Native in the jungles of Assam & Kerala, Burma, Ceylon. Common in Orissa, Bengal South India & Assam.

Taxonomic Description: A deciduous tree, with large showy flowers and woody capsule fruits.

Medicinal Parts: Fruits, Seeds, Leaves, Bark, Root.

Chemical Constituents: Leaves - Alanine, Isoleucine, a-aminobutyric acid, Menthoeonine, Lageracetal, Amylalcohol, Ellagic acid, β-sitosterol, lager tannin.

Fruit, Bark - Tannin.

Medicinal Properties:

तिनिशः श्लेष्मपित्तास्त्रमेदः कुष्ठप्रमेहजित् ।
तुवरः श्वित्रदाहघ्नो व्रणपाण्डुक्रिमिप्रणुत् ।। (भा. प्र.)

Tiniśaḥ śleṣmapiṭtāṣṭramedaḥ kuṣtha pramehajiṭ
Tuvaraḥ ṣviṭradāhaghno vraṇapāṇdukṛimipṛaṇuṭ. (Bhā. Pṛa.)

Root - Astringent, Febrifug, Stimulant.

Bark / Leaves - Purgative.

Seeds - Narcotic.

Medicinal Uses: Bark - Fever.

Ethno- Medicinal Uses: Bark - Dysentery (Banerjee, 2003).

LAUNEA COROMANDELICA (Houtt.) Merr. (ANACARDIACEAE)

Syn.: *Odina wodier* Roxb.

Regional & Other Names : *Bengali* – Jiol; *English* – Wodier, Jhingam; *Gujarati* - Jinni; *Hindi* – Jhingan; *Kannada* – Manjistha; *Malayalam* - Karasu; *Marathi* – Mai; *Sanskrit* – Jhingini; *Tamil* – Oti; *Telugu* – Oddimanu.

Distribution: Throughout India.

Taxonomic Description: Medium sized trees, with Yellowsih to Purplish flowers and drupes fruits.

Medicinal Parts: Bark, Leaves.

Chemical Constituents: Bark - β-sitosterol, Physcion, Phlobatannins, Physcion anthranol B.

Leaves - Ellagic acid, Quercetin, β-sitosterol, Rutin, Leucocyanidin, Leucodelphinidin.

Flower - Quercetin, Iso-quercitrin, Ellagic acid, Morin.

Medicinal Properties:

जिङ्.गिणी व्रणहृद्रोगवातातीसारजित्कटुः ।
उष्णास्तिस्यास्तु निर्यासो नस्याद्बाहुव्यथापहः ।। (म. वि.)

Jiṇgiṇī vṛaṇahṛdrogavātātīsārajiṭkatuḥ
Uṣṇāṣtasyāṣtu niṛyāso naṣyāḍbāhuvyathāpahaḥ. (Ma. Vi.)

Bark - Acrid, Anodyne, Astringent, Stomachic, Sweet, Thermogenic.

Medicinal Uses: Bark - Bruises, Cuts, Diarrhoea, Dysentery.

Gout, Sprains, Ulcerative Stomatitis, Ulcers, Wounds.

Leaves – Bruises, Elephantiasis, Inflammation, Neuralgia, Sprains.

LEEA INDICA (Burm. f.) Merr. (LEEACEAE)

Syn.: *L. Sambucina* (L.) Willd.

Regional & Other Names : *Bengali* – Kurkurjiwah; *Hindi* – Kurkur-jihva; *Kannada* – Gadhapatri; *Malayalam* - Naluku; *Marathi* – Oino; *Sanskrit* – Chatri; *Tamil* – Naikki; *Telugu* – Ankadora.

Distribution: Throughout India.

Taxonomic Description: Large shrubs, with corymbose cyme inflorescence and Berries fruits.

Medicinal Parts: Root, Leaves.

Chemical Constituents: Leaves contain amorphous froth forming acid.

Medicinal Properties:

कुक्कुरजिह्वा कषायकटुका नात्युष्णा कृमिशोफनुत् ।। (स्व.)

"Kukkurajiḥvā kasāyakatukā nāṭyuṣṇā kṛmiśophanuṭ." (Ṣva.)

Root - Anthelmintic, Astringent, Acrid, Cooling, Sudorific.

Medicinal Uses: Root - Colic, Diarrhoea, Dysentery, Hyperdipsia, Skin diseases, Ulcers.

Ethno- Medicinal Uses: Root - Diarrhoea, Dysentery (Sinha *et al.,* 2006; Shah *et al.,* 1981, 1983).

LEPTADENIA RETICULATA Wight & Arn. (ASCLEPIADACEAE)

Regional & Other Names : *Gujarati* – Dodi; *Hindi* – Dori; *Marathi* – Dodi; *Sanskrit* – Jivanti, Meda; *Tamil* – Palaikkodi; *Telugu* – Kalasa.

Distribution: Punjab, Western peninsula, Ceylon, Burma, Malay Peninsula.

Taxonomic Description: Perennial climbers, with green or yellow flowers and follicle fruits.

Medicinal Parts : Root.

Medicinal Properties:

जीवन्ती शीतला स्वादुः स्निग्धा दोषत्रयापहा ।
रसायनी बलकरी चक्षुष्या ग्राहिणी लघुः ।। (भा. प्र.)

Jīvantī śītalā svāduḥ sṇigdhā dosatryāpahā
Rasāyanī balkarī cakshusyā grāhinī laghuḥ (Bha. Pra.)

Stimulant, Restorative.

Medicinal Uses: Leaves/Roots - Skin affections.

Plant - Habitual abortion.

Ethno- Medicinal Uses: Leaf - Cough, Skin diseases (Shah et al., 1983); Asthma (Sharma et al., 1979; Sinha *et al.,* 2006); skin diseases (Sinha *et al.,* 2006).

Whole plant - Ear & Nose complaint (Sharma *et al.,* 1979); Tonic (Sudhakar & Rolla, 1985).

LILIUM POLYPHYLLUM D. Don (LILIACEAE)

Regional & Other Names : *English* – Ksheerakakoli; *Gujarati* - Kakoli; *Hindi* – Ksirakakoli; *Kannada* – Ksirakakoli; *Malayalam* - Ksirakakoli; *Marathi* – Kakoli; *Sanskrit* – Ksirakakoli; *Tamil* – Ksirakakoli; *Telugu* – Ksirakakoli.

Distribution: The plant in found in Western temperate Himalaya from Kumayun to Kashmir from 1800-3600 m. altitude.

Taxonomic Description: Bulbous herbs, with fragrant, dull Yellowish pendulous flowers & capsule fruits.

Medicinal Parts: Bulb.

Chemical Constituents: Tubers - Linalool, □-terpineol.

Medicinal Properties:

काकोलीयुगलं शीतं शुक्ललं मधुरं गुरू ।
बृहणं वातदाहास्रपित्तशोफज्वरापहम् ।। (भा. प्र.)

Kākolīyugalaṃ śītaṃ śuklalaṃ madhuraṃ gurū
Bṛmhaṇaṃ vātadāhāsṛapiṭtaśophajvarāpahaṃ. (Bhā. Pṛa.)

Bulb - Aphrodisiac, Bitter, Diuretic, Expectorant, Antipyretic, Refrigerant, Sweet, Tonic.

Medicinal Uses: Bulb - Bronchitis, Burning sensation, Cough, General debility, Hyperdipsia, Intermittent fever, Seminal weakness, Strangury.

Ethno- Medicinal Uses: Bulb - Tonic (Uniyal, 1968).

LIMNOPHILLA AROMATICA (L.) Merr. (SCROPHULARIACEAE)

Syn.: *L. gratissima* Blume.

Regional & Other Names : *Bengali* –Karpur; *Hindi* – Kutra; *Kannada* – Mangannari; *Malayalam* - Mannanari; *Marathi* – Ambuli; *Sanskrit* – Amragandhah; *Tamil* – Mankainari.

Distribution: Throughout India.

Taxonomic Description: An aromatic herb, with Purplish flowers and capsule fruit.

Medicinal Parts : Whole plant.

Chemical Constituents: The plant contains d-limonene, d-perillaldehyde, a flavone as 7-desmethyl artemetin.

Medicinal Properties:

आम्रगंधा दीपनी पाचनी शीतला च रूचिप्रदा ।। (स्व.)

"Āmṛagaṇdhā dīpanī pācani śītalā ca rūcipṛadā." (Ṣva.)

Whole plant - Anthlemintic, Anti-inflammatory, Antiseptic, Aperient, Appetiser, Carminative, Digestive, Diuretic, Emollient, Febrifuge, Refrigerant, Slightly bitter, Sour.

Medicinal Uses: Whole plant - Agalactia, Anorexia, Constipation, Dyspepsia, Fever, Helminthiasis, Inflammation, Strangury.

Ethno- Medicinal Uses: Whole plant - Dysentery (Mukhopadhyay, 1992).

LIQUIDAMBER ORIENTALIS Mill. (HAMAMELIDACEAE)

Regional & Other Names : *Bengali* –Silaras; *English* – Oriental sweet-gum; *Gujarati* – Silaras; *Hindi* – Silaras; *Kannada* – Kundurka; *Malayalam* - Arabikkunturukkam; *Marathi* –Silaras; *Sanskrit* – Sihlakah; *Tamil* – Neri-arisippal; *Telugu* – Silarasu.

Distribution: In Asia Minor.

Taxonomic Description: Medium sized trees, with Palmately 5-lobed leaves & Yellow monoecious flowers.

Medicinal Parts: Balsam.

Chemical Contituents: Balsam - Amorphous substance storesin, cinnamic acid.

Medicinal Properties:

स्निग्धः सुगन्धोऽनिलहा कायवैभत्स्यनाशनः ।
पिटकाकोठशमनस्तुरूष्कः कफपित्तजित् ।। (म. नि.)

Śnigdhaḥ sugandhoanilahā kāyavaībhaṭṣyanāśanaḥ
Pitakākothaśamanaṣturuṣkaḥ kaphapiṭtajiṭ. (Ma. Ni.)

Balsam - Acrid, Anodyne, Antibacterial, Aphrodisiac, Aromatic, Bitter, Constipating, Deodrant, Emollient, Expectorant, Depurative, Diuretic, Febrifuge, Sweet, Thermogenic, Tonic. Vulnerary.

Medicinal Uses: Balsam -, Amenorrhoea, Bronchitis, Chronic cough, Diarrhoea, Fever, Foul ulcers, General debility, Inflammation, Leprosy, Pruritus, Skin diseases, Strangury, Wounds.

MACROTYLOMA UNIFLORUM (Lam.) Verdc. (FABACEAE)

Syn.: *Dolichos biflorus* L.

Regional & Other Names : *English* – Horse gram; *Gujarati* - Kulathi; *Hindi* – Kulthi; *Kannada* – Huruli; *Malayalam* - Mutira; *Marathi* – Kulitha; *Sanskrit* – Kutatthah; *Tamil* – Kollu; *Telugu* – Ulavalu.

Distribution: Cultivated throughout India.

Taxonomic Description: Trailing annual plants, with Yellow flowers & Pods fruits.

Medicinal Parts: Seeds.

Chemical Constituents: Seed contains - Urease, Strepogenin, β-sitosterol, Genistein, Kievitone, Phaseollidin, Isoferreirin, Coumesterol, Psoralidin, d-glucose, d-galactose, L-rhamnose, d-arabinose, L-ascorbic acid, Amino acids-glycine, Alanine, Cysteine, Serine, Aspartic acid.

Medicinal Properties:

ऊष्णाः कषायाः पाकेऽम्लाः कफशुक्लानिलापहः ।
कुलत्थाः ग्राहिणाः कासहिक्काश्वासर्शसां हिताः ।। (च. सं.)

Ūṣṇāḥ kasāyāḥ pākeamlāḥ kaphaśuklānilāpahaḥ
Kulatthāḥ grāhiṇāḥ kāsahikkāṣvāsarṣasāṃ hitāḥ. (Ca. Saṃ.)

Seed - Acrid, Anthlemintic, Astringent, Bitter, Diaphoretic, Diuretic, Expectorant, Febrifuge, Ophthalmic, Thermogenic, Tonic.

Medicinal Uses: Seed - Asthma, Bronchitis, Hiccough, Tumours, Strangury, Verminosis.

Ethno- Medicinal Uses: Seed - Kidney stone (Jain,1984; Lal & Yadav, 1983; Sinha *et al.,* 2006); Menstrual trouble (Abraham, 1981); Leprosy, Toothache (Sinha *et al.,* 2006).

MALAXIS MUSCIFERA (Lindley) Kuntze. (ORCHIDACEAE)

Syn.: *Microstylis muscifera* (Lindley) Ridley.

Regional & Other Names : *English* – Risabak; *Hindi* – Rsabhak; *Kannada* – Rsabaka; *Malayalam* - Itavakam; *Sanskrit* – Rsabhakah; *Tamil* – Rsabakam; *Telugu* – Rsabaka.

Distribution: Throughout India.

Taxonomic Description: Terrestrial herbs with small ovoid Pseudo-bulb and Yellowish-Green flowers.

Medicinal Parts: Green swollen stem base covered with Brown scales.

Medicinal Properties:

जीवकर्षभकौ शीतौ बृहणौ कफशुक्ललौ ।
मधुरो दाहपित्तास्रदाहक्षयनिबर्हणौ ।। (कै. नि.)

Jīvakarsbhakaū śītaū bṛṃhaṇaū kaphaṣuklalaū
Madhurō dāhapiṭtāsṛadāhasayanibaṛhaṇaū. (Kaī. Ni.)

Swollen stem - Antidysenteric, Aphrodisiac, Febrifuge, Refrigerant, Styptic, Sweet, Tonic.

Medicinal Uses: Swollen stem - Burning sensation, Dysentery, Fever, General debility, Haemorrhages, Seminal weakness, Sterility.

MANIHOT ESCULENTA Crantz. (EUPHORBIACEAE)

Syn.: *M. utilissima* Pohl.; *M. palmata* Muell.-Arg.; *M. aipi* Phol.; *M. dulcis* Pax.

Regional & Other Names : *English* – Tapioca; *Hindi* – Sakkarkand; *Kannada* – Maragenasu; *Malayalam* - Maraccini; *Sanskrit* – Kalapakandah; *Tamil* – Maravallikkilanku; *Telugu* – Karrapendalamu..

Distribution: Cultivated widely in Kerala and Tamil Nadu.

Taxonomic Description: A shrub, with palmately compound leaves and Yellow or Greenish White unisexual flowers.

Medicinal Parts: Tuberous roots.

Chemical Constituents: Plant contains Cyanogenetic Glucose. Juice is poisonous.

Medicinal Properties:

दारूकन्दः रूच्यो बल्यो मधुरो रसपाकयोः ।। (स्व.)

"Dārūkaṇdaḥ rūcyo baḷyo madhurō raspākayoḥ." (Ṣva.)

Tuberous roots - Aperient, Appetiser, Nourishing, Sweet, Tonic, Vulnerary.

Medicinal Uses: Tuberous roots - Anorexia, Constipation, Dyspepsia, Foul ulcers, General debility, Wounds.

Ethno- Medicinal Uses: Leaf - Skin diseases (Rao & Jamir, 1982).

Root - Stomach pain, Chest pain, Diarrhoea (Rajan *et al.,* 2003).

MARANTA ARUNDINACEA L. (MARANTACEAE)

Regional & Other Names : *Bengali* – Ararut; *English* – Arrowroot; *Gujarati* –Ararat; *Hindi*– Tikhor; *Kannada* – Kuvehittu; *Malayalam* - Kuvva; *Marathi*–Tavkil; *Sanskrit*–Tavaksiri; *Tamil*–Kuvai; *Telugu*– Palagunta.

Distribution: Cultivated throughout India.

Taxonomic Description: Slender erect rhizomatous herbs, with ovate-lanceolate leaves, White flowers.

Medicinal Parts: Rhizome.

Chemical Constituents: Tuber contains - Starch, Thin skin of tubers contains Bitter principles.

Medicinal Properties:

तवक्षीरं तु मधुरं शिशिरं दाहपित्तनुत् ।
क्षयकासकफश्वासनाशनं चास्रदोषनुत् ।। (रा. नि.)

Tavaksīraṃ tu madhuraṃ śiśiraṃ dāhapiṭtanuṭ
Kasayakāsakaphaṣvāsanāśanaṃ cāsṛadosanuṭ. (Rā. Ni.)

Rhizome - Aphrodisiac, Astringent, Emollient, Expectorant, Febrifuge, Refrigerant, Rubefacient, Sweet, Tonic.

Medicinal Uses: Rhizome - Bronchitis, Cough, Diarrhoea, Dysentery, Dyspepsia.

MARTYNIA ANNUA L. (MARTYNIACEAE)

Regional & Other Names : *Bengali* – Bagh noki; *English* – Devil's claw; *Gujarati* –Vichchida; *Hindi* – Bichua; *Malayalam* - Winchauri; *Marathi* – Vinchu; *Sanskrit* – Kaknasa; *Tamil* – Nagathali; *Telugu* – Garudamukku.

Distribution: Native of mexico, becoming naturalized in India.

Taxonomic Description: Glandular pubescent herbs, with fistular stem, ovate-cordate leaves & drooping, pinkish-white flowers. Fruits woody, beetle-shaped, hooked at the tip.

Medicinal Parts: Fruit, Leaves.

Chemical Constituents: Leaves -Chlorogenic acid.

Seeds - Pale Yellow semi drying oil, Fatty oil: Fatty acids of the fatty oil are Palmitic, Stearic, arachidic, Oleic acid, Linoleic etc.

Fruits - Fatty oil: Fatty acids of the fatty oil are Plamitic, Stearic, Oleic acid, Linoleic acid.

Medicinal Properties:

काकनासा कषायोष्णा कटुका रसपाकयोः ।
कफघ्नी वामनी तिक्ता शोथार्शश्वित्रकुष्ठहृत् ।। (भा. प्र.)

Kāknāsā kaṣāyōṣanā katukā rasapākayoḥ
Kaphaghanī vāmanī tiktā sodhārsasvitrakusthahṛt (Bha. Pra.)

Fruits – Alexiteric, Anti-Inflammatory.

Medicinal Uses: Fruits - Inflammation.

Leaves - Epilepsy, Sore throat.

Ethno- Medicinal Uses: Seed - Eczema (Maheshwari *et al.,* 1986; Sinha *et al.,* 2006); Sore (Jain & De, 1966); Rheumatic pain, Epilepsy (Sinha *et al.,* 2006).

Leaf - Epilepsy (Sudhakar & Rolla, 1985; Shah *et al.,* 1983; Mukhopadhyay, 1992; Sinha *et al.,* 2006); Sore throat (Sudhakar & Rolla,

1985; Bedi, 1978); Leprosy (Deokule, 2006); Eczema, Rheumatic pain (Sinha *et al.*, 2006).

Fruits - Itching (Saxena & Vyas, 1983); Rheumatism (Tribedi *et al.*, 1982); Scabies (Singh *et al.*, 1992).

Root- Tonsilitis (Joshi *et al.*, 1980).

Seed oil - Skin diseases (Shukla *et al.*, 1992); Eczema, Boils (Girach *et al.*, 2003; Shukla *et al.*, 1992).

MENTHA SPICATA L. (LAMIACEAE)

Syn.: *Mentha vidrdis* L.

Regional & Other Names : *Bengali* – Pudina; *English* – Garden mint; *Hindi* – Podina; *Kannada* – Pudina; *Malayalam* - Putiyina; *Marathi* – Pudina; *Sanskrit*-Putiha; *Tamil* – Jach-chak; *Telugu* – Pudina.

Distribution: Northern temperate regions.

Taxonomic Description: Aromatic, creeping or ascending herbs. Leaves ovate or ovate-elliptic. Flowers purplish.

Medicinal Parts: Leaves.

Chemical Constituents: Diosmetin, Diosmin from leaves. Essential oil contains Carvacrol, L-carvone, Dihydrocarveol, Limonene, Dipentene and Esters.

Medicinal Properties:

पूतिहा कटुरूष्णश्च रोचनी दीपनो लघुः ।
हन्ति वात कफाध्मानशूलच्छर्दिकृमीस्तथा ।।

Pūtihā katurūṣṇaṣyca rōcanī dīpno laghuḥ
Haṇti vātakaphādhmānaśūlacchaṛdikṛmīmṣtathā.

Antispasmodic, Carminative, Stimulant.

Medicinal Uses: Leaves – Bronchtis, Fever.

Ethno- Medicinal Uses: Leaf - Vomiting (Satapathy & Brahman, 2003); Diarrhoea (Jadeja *et al.*, 2006).

Whole plant - Stimulant, Antispasmodic (Basak, 2006).

MERREMIA TRIDENTATA (L.) Hallier f ssp. *TRIDENTATA* (CONVOLVULACEAE)

Syn.: *Ipomoea tridentata* (L.) Roth.

Regional & Other Names : *Malayalam* - Prasarini; *Sanskrit* – Prasarani; *Tamil* – Mutiyakuntal; *Telugu* – Sitasavaram.

Distribution: Throughout India.

Taxonomic Description: A small glabrous perennial herb with many decumbent stems from a small woody rootstock. Leaves sessile. Flowers axillary, peduncles wiry, 1-3-flowered; pedicels, sub-clavate. Sepals elliptic, glabrous. Capsules ovoid, glabrous, 2-celled. Seeds dark, glabrous.

Medicinal Parts: Whole plant.

Chemical Constituents: Diosmetin, luteolin and their 7-O-glucosides.

Medicinal Properties:

प्रसारणी गुरूष्णा च तिक्ता वातविनाशिनी ।
अर्शःश्वयथुहन्त्री च मलविष्टम्भहारिणी ।। (रा. नि.)

Pṛasāraṇī guṛuṣṇā ca tiktā vātavināṣinī
Aṛṣaḥṣvayathuhaṇṭṛī ca malaviṣtaṃbhahāriṇī. (Rā. Ni.)

Anodyne, Aphrodisiac, Astringent, Bitter, Laxative, Thermogenic, Tonic.

Medicinal Uses: Plant - General debility, Haemorrhoids, Inflammation, Inflammation,Uropathy.

Ethno- Medicinal Uses: Shoot - Wounds (Tarafder, 1984).

Root - Body pain (Sur *et al.*, 1992).

MONOCHORIA VAGINALIS Presl (PONTEDERIACEAE)

Regional & Other Names : *Bengali*-Nukha; *Malayalam* - Karin-kuvvalam; *Sanskrit* – Indivarah; *Tamil* – Karimkuvalam; *Telugu* – Nirkanca.

Distribution: Throughout India.

Taxonomic Description: Aquatic perennial herbs, with Blue flowers and ellipsoid capsules fruits.

Medicinal Parts: Root stock, Leaves.

Medicinal Properties:

नीलोत्पलमतिस्वादु शीतं सुरभि सौख्यकृत् ।
पाके तु तिक्तमत्यन्तं रक्तपित्तापहारकम् ।। (रा. नि.)

Nīlolpaḷamatiṣvādu śītaṃ surabhi saūkhyakṛt
Pāke tu tiktamaṭyaṇtaṃ raktapiṭtāpahārakaṃ. (Rā. Ni.)

Root stock / Leaves - Alterant, Aromatic, Bitter, Cooling, Diuretic, Sweet, Tonic.

Medicinal Uses: Root stock / Leaves - Asthma, Boils, Cough, Dipsia, Gastropathy, Haemorrhages. Hepatopathy, Scurvy, Strangury,

Ethno- Medicinal Uses: Px - Anaemia, Dysentery, Rheumatism (Jain & Tarafder, 1970).

MORINDA PUBESCENS J.E. Smith. (RUBIACEAE)

Syn.: *M. coreia* Buch-Ham.; *M. tinctoria* Roxb.

Regional & Other Names : *English* – Morinda tree; *Hindi* – Acc; *Kannada* – Haladipavete; *Malayalam* - Pavitta; *Sanskrit* – Paphanah; *Tamil* – Mannanatti; *Telugu* – Maddicettu.

Distribution: Bengal, Bihar, Central Provinces & South India.

Taxonomic Description: Moderate sized deciduous trees, with White flowers & drupe fruits.

Medicinal Parts: Roots, Leaves.

Chemical Constituents: Root - Morindone and its Glucoside Morindine; Damnacanthal, Nordamnacanthal, β-Sitosterol, Anthragallol-2-3-dimethyl ether, Soranjidol, Ibericin, Morindone-6-Primeveroside.

Leaves - Campesterol, Stigmasterol, β-sitosterol, Ursolic acid.

Medicinal Properties:

मिथ्योगुच्छः पफणश्च वृकधूमो निगद्यते ।। (अ. म.)

Mithyogucchaḥ paphaṇaṣca ṿrkadhūmo nigaḍyate. (A. Ma.)

Root - Styptic, Constipating, Anti-inflammatory, Alexeteric, Tonic.

Leaves - Digestive, Carminative, Febrifuge, tonic.

Medicinal Uses: Root - Boils, Dysentery, General debility, Haemorrhages, Inflammation.

Leaves - Diarrhoea, Dyspepsia, Gaotropathy, Gout, Inflammation,

Ulcerative stomatitis, Wounds.

Ethno- Medicinal Uses: Root - Ulcer (Tribedi *et al.,* 1982).

Bark - Urinary complaint (Jain & De, 1966).

Leaves - Stomach pain & dysentery in children (Ganesan & Kesavan, 2003); Wounds (Arinathan *et al.,* 2006).

MUSSAENDA FRONDOSA L. (RUBIACEAE)

Regional & Other Names : *Bengali* – Nagballi; *English* – White calycine; *Hindi* – Bedina; *Kannada* – Belloti; *Malayalam* - Vellila; *Marathi* – Bhurtkasi; *Sanskrit* – Rajatarih; *Tamil* – Vellai ilai; *Telugu* – Nagavalli.

Distribution: Throughout India.

Taxonomic Description: Erect or scandent shrub, with Yellowish Green outside and Orange Green inside flowers and Berries fruits.

Medicinal Parts: Whole plant.

Chemical Constituents: Sepals - Ferulic acid, Hyperin, Quercetin, Rutin, Sinapic acid, β -sitosterol glucoside.

Medicinal Properties:

रजतारिः दृष्टिशुक्लं जयेत् रक्तोत्भवारूजा ।। (ह्र. प्रि.)

Rajatāriḥ dṛṣtiśuklaṃ jayeṭ raktoṭbhavārūjah. (Hṛ. Pṛi.)

Alterant, Anti-Inflammatory, Astringent, Demulcent, Expectorant, Febrifuge, Cardiotonic, Ophthalmic, Sweet, Vulnerary.

Medicinal Uses: Plant - Bronchitis, Cough, Fever, Inflammation, Jaundice, Leucoderma, Ophthalmopathy, Pruritus, Ulcers, Wounds.

Ethno- Medicinal Uses: Flower – Asthma, Dropsy, Fever (Mukherjee *et al.,* 1986).

Root - Fever (Krishna & Singh, 1987); Leucoderma (Baruah *et al.*, 1987); Leprosy (Mukherjee *et al.*, 1986).

Leaf - Swelling (Karkongor & Joseph, 1981).

Sepals - Jaundice (Mukherjee *et al.*, 1986).

MYXOPYRUM SERRATULUM A. W. Hill. (OLEACEAE)

Syn.: *M. smilacifolium* C. B. Clarke

Regional & Other Names : *Hindi* – Pilacameli; *Malayalam* - Caturamulla; *Sanskrit* – Hemamalati; *Tamil* – Caturamullai; *Telugu* – Caturdharalata.

Distribution: Throughout Kerala, in evergreen forests at altitude of 600 to 900 m.

Taxonomic Description: Scandent large shrubs, with Yellowish small flowers and Berries fruits.

Medicinal Parts: Root, Leaves.

Medicinal Properties:

हेममालती कटुकोष्णा मधुरा कषायकः ।। (स्व.)

Hemamālatī katukoṣṇā madhurā kasāyakaḥ. (Ṣva.)

Leaves -Acrid, Anodyne, Astringent, Febrifuge, Sweet, Thermogenic, Tonic.

Medicinal Uses: Root - Scabies.

Leaves - Asthma, Cough, Cuts, Fever, Rheumatism, Wounds.

Ethno- Medicinal Uses: Leaves - Asthma, Cough, Rehumatism (Nadanakunjidam, 2003).

NARAVELIA ZEYLANICA (L.) DC. (RANUNCULACEAE)

Regional & Other Names: *Bengali* – Chagul-bati; *Kannada* –Neenda-malli; *Malayalam* - Poytalacci; *Sanskrit* – Dhanavalli; *Tamil* – Vatamkolli .

Distribution: Throughout India.

Taxonomic Description: Climbing tendriller shrubs, with fragrant Yellow flowers & aggregate of achenes fruits.

Medicinal Parts: Whole plant.

Medicinal Properties:

धनवल्ली मधुरा तिक्ता कषाया वातशुलनुत् ।। (स्व.)

Dhanavaḷlī madhurā tiktā kasāyā vātaśulnuṭ. (Ṣva.)

Anodyne, Anthelmintic, Anti-inflammatory Bitter, Depurative, Sweet, Vulnerary.

Medicinal Uses: Colic, Helminthiasis, Inflammation, Leprosy, Ulcers, Wounds.

Ethno- Medicinal Uses: Leaf – Cold (Ramachandran & Nair, 1981); Toothache (Gantait *et al.,* 2006); Headache (Ramachandran & Nair, 1981; Gantait *et al.,* 2006).

Whole plant -Body pain (Sasidharan & Augustine, 2006).

Root - Toothache (Sinha *et al.,* 2006); Fever (Udayan, *et al.,* 2006); Cough (Gantait *et al.,* 2006; Sinha *et al.,* 2006); Cold, (Sinha *et al.,* 2006; Udayan, *et al.,* 2006).

NERVILIA ARAGOANA Gaud. (ORCHIDACEAE)

Syn.: *Pogonia flabelliformis* Lindl.; *P. carinata* Wt.

Regional & Other Names : Hindi – Sthalapadma; Malayalam - Kalttamara; Sanskrit – Padmacarini; Tamil – Orilaittamarai.

Distribution: In the Travancore & Pulney hills at all elevations.

Taxonomic Description: Terrestrial orchid with a single lotus like leaf, sepals & petals Pale-Green veined with Purple, lip White 3-lobed about the middle, veined with Purple, tinged with Yellow near the base, mid lobe ovate, acute, Crisped, Hairy.

Medicinal Parts: Whole plant.

Chemical Constituents: Phytol, cycloeucalenol, stigmasterol, linoleic acid, linolenic acid, L-norleucine.

Medicinal Properties:

पद्मानुष्णा कटुस्तिक्ता कषाया कफवातजित् ।
मूत्रकृच्छ्राश्मशूलघ्नी श्वासकासविषापहा ।। (भा. प्र.)

Paḍmānuṣṇā katuṣtiktā kasāyā kaphavātajiṭ
Mūṭrakrcchrāṣmaśūlaghnī ṣvāsakāsavisāpahā. (Bhā. Pṛa.)

The plant is Bitter, Acrid, Cooling, Diuretic, Tonic.

Medicinal Uses: Agalactia, Asthma, Colic, Cough, Diarrhoea, Haemoptysis, Mental instability, Vomiting.

Ethno- Medicinal Uses: Tuber - Cooling (Saxena, 1986).

Root tuber - To improve body vigour (Udayan, *et al.,* 2006).

NILGIRIANTHUS CILIATUS (Nees.) Bremek. (ACANTHACEAE)

Syn.: *Strobilanthues ciliatus* Nees.

Regional & Other Names : *Hindi* – Karvi; *Malayalam* - Kurunni; *Sanskrit* – Sahacarah; *Tamil* – Kurunji.

Distribution: Throughout the evergreen forests of Western Ghats upto 1,200 m.

Taxonomic Description: Slender shrubs, with subquadrangular stems and White to liliac fruits.

Medicinal Parts: Root, Leaves, Bark.

Medicinal Properties:

सैरेयः कुष्ठवातास्रकफकण्डूविषापहः ।
तिक्तोष्णो मधुरो दन्त्यः सुस्निग्धः केशरंजनः ।। (भा. प्र.)

Saīreyaḥ kuṣthavātāsṛakaphakaṇdūvisāpahaḥ
Tiktoṣṇo madhurō daṇtyaḥ suṣnigdhaḥ keṣaraṇjanaḥ.(Bhā. Pṛa.)

Root - Anti-Inflammatory, Bitter, Depurative, Diaphoretic, Diuretic, Emollient, Expectorant, Febrifuge, Sweet, Thermogenic.

Leaf / Bark - Depurative, Diaphoretic, Expectorant, Febirfuge.

Medicinal Uses: Root – Bronchitis, Chest congestion, Cough, Fever, Inflammation, Leucoderma, Lumbago, Sciatica, Skin diseases, Strangury.

Leaf/Bark - Bronchitis, Dropsy, Fever, Inflammation, Leprosy, Leucoderma, Pruritus, Scrofula, Whooping cough.

Ethno- Medicinal Uses: Whole plant - Tooth problem (Thomas & Britto, 2003)

OCIMUM KILIMANDSCHARICUM Guerke. (LAMIACEAE)

Regional & Other Names : *Bengali* –Kapur tulsi; *English* – Camphor basil; *Hindi* – Kapurtulsi; *Kannada* – Karpuratulasi; *Malayalam* - Karppura-ttulasi; *Sanskrit* – Karpuratulasi; *Tamil* – Karpuratulasi; *Telugu* – Karpura-tulasi.

Distribution: Throughout India.

Taxonomic Description: Aromatic undeshrubs with pubescent quadrangular branchlets; leaves simple, opposite, oblong, narrow at the base, deeply serrate; flowers in 4-6 flowered whorls with a densely pubescent axis, corolla 2-3 times the length of the calyx; fruits ovoid-oblong nultlets.

Medicinal Parts: Leaves.

Chemical Constituents: Leaves yield an essential oil 6-13% of which 50-75% is camphor.

Medicinal Properties:

भूतावेश्या लता चान्या कर्पूरतुलसी स्मृता ।। (कै. नि.)

Bhūtāveṣyā latā cāṇyā kaṛpūratulasī ṣmṛtā. (Kaī. Ni.)

Leaves - Acrid, Anti-bacterial, Antiviral, Appetising, Aromatic, Insecticidal, Ophthalmic, Thermogenic,

Medicinal Uses: Leaves - Anorexia, Bacterial & Viral infections, Bronchitis, Catarrh, Cough, Foul ulcers, Ophthalmopathy, Wounds.

Ethno- Medicinal Uses: Leaves - Fever, Cold, Cough (Sood & Thakur, 2004).

OPHIORRHIZA MUNGOS L. (RUBIACEAE)

Regional & Other Names : *Bengali* – Gandhanakuli; *English* – Mongoose plant; *Gujarati* – Mungusvel; *Hindi* – Sarhati; *Kannada* – Patalagaruda; *Malayalam* - Avilpori; *Marathi* – Mungusavela; *Sanskrit* – Sarpari; *Tamil* – Kirippuntu; *Telugu* – Sarpasicettu.

Distribution: Throughout India.

Taxonomic Description: A herbaceous undershrubs, with White flowers & capsule fruits.

Medicinal Parts: Roots.

Chemical Constituents-: Root - β-sitosterol, 5a-ergost-7-en-3β-ol and 5□-ergost-8(14)-en-3β-ol.

Medicinal Properties:

सर्पाक्षी कटुका तिक्ता सोष्णा कृमिनिकृन्तनी ।
वृश्चीकोन्दुरूसर्पानां विषघ्नी व्रणरोपणी ।। (भा. प्र.)

Saṛpāksī katukā tiktā soṣṇā kṛminikṛṇtanī
Vṛṣcīkoṇdurusaṛ pānāṃ visaghnī vṛanarōpaṇī. (Bhā. Pṛa.)

Root: Acrid, Alexeteric, Anthelmintic, Bitter, Laxative, Sedative, Stomachic, Thermogenic, Tonic, Vulnerary,

Medicinal Uses: Root - Cancer, Gastropathy, Helminthiasis, Hydrophobia, Leprosy, Ulcers, Wounds.

Ethno- Medicinal Uses: Whole plant - Stomachache (Mukherjee *et al.*, 1986).

OPUNTIA ELATIOR Mill. (CACTACEAE)

Syn.: *O. nigricans* Haw.

Regional & Other Names : *English* – Prickly pear; *Hindi* – Naagaphanaa; *Sanskrit* – Kantaseva.

Distribution: Throughout India.

Taxonomic Description: Spreading bush, with flat, succulent, jointed, obovate branches having several straight spines on each cushion.

Medicinal Parts: Fruits, Leaves, Latex.

Chemical Constituents: Flower contains Glycoside of Isorhamnetin and Quercertain and Glucoside narcissi. Isolation of Isoquercitrin, β-sitosterol, Opintiol. A Polysaccharide contains Galactose and Arabinose from Pods.

Medicinal Properties:

कन्थारी कटु तिक्तोष्णा कफवातनिकृन्तनी ।
शोफघ्नी दीपनी रूच्या रक्तग्रन्थीरूजापहा ।। (रा. नि.)

Kaṇthārī katu tiktoṣṇā kaphavātanikṛṇtanī
Śophaghnī dīpanī rūcyā raktagraṇthīrūjāpahā. (Rā. Ni.)

Fruits - Refrigerant.

Latex - Purgative.

Medicinal Uses: Fruits - Gonorrhoea, Whooping Cough.

Ethno- Medicinal Uses: Stem – Boils (Shekhawat & Sunil Anand, 1984); burns, muscular swelling, wounds (Shah *et al.,* 1983); Boils (Shukla *et al.,* 1992).

Phylloclade - Painful swellings & Sprains (Kumar & Chauhan,2005).

Leaf - Poultice is applied for inflammed parts (Deokule, 2006).

Fruit juice - Children's cough (Deokule, 2006).

Leaf, Fruit - Gonorrhoea, Purgative (Sinha *et al.,* 2006).

ORCHIS LATIFOLIA L. (ORCHIDACEAE)

Syn.: *Dactylorhiza hatagirea* (Don.) Soo.

Regional & Other Names : *English* – Marsh orchid; *Hindi* – Salam; *Kannada* – Salamisri; *Malayalam* - Salamisri; *Sanskrit* – Salampamisri; *Tamil* – Salamisri; *Telugu* – Salamisri.

Distribution: In the Himalayas, at an altitude of 2,500-5,000 m.

Taxonomic Description: Tuberous terrestrial orchid, with Yellowish, White or Greyish tubers & Dull Purple flowers.

Medicinal Parts: Tubers.

Chemical Constituents: Tuber contains - A Glucoside, a Bitter Substance, Starch, Mucilage, Sugar, Albumen, Volatile oil; Phosphates & Chlorides of Potassium and Lime; Starch.

Medicinal Properties:

बल्यः शीतो गुरूः स्निग्धस्तर्पणो बृंहणात्मकः ।
वातपित्तहरः स्वादुर्वृष्यो मुञ्जातकः परम् ।। (च. सू.)

Bạlyaḥ śīto gurūḥ ṣnigdhaṣtaṛpaṇo ḅrṃhaṇāṭmakaḥ
Vātapiṭtaharaḥ ṣvāduṛvṛṣyo muṇjātakaḥ paraṃ. (Ca. Sū.)

Tubers - Aphrodisiac, Cooling, Emollient, Rejuvenating, Sweet, Tonic.

Medicinal Uses: Tubers - Chronic Diarrhoea, Diabetes, Dysentery, General debility, Seminal weakness.

Ethno- Medicinal Uses: Tuber – Astringent (Gaur *et al.,* 1983); Bone fracture (Tiwari *et al.,* 1980); Tonic (Uniyal, 1968; Gaur *et al.,* 1983; Gupta, 1981); Wounds (Bennet, 1985; Kapur & Nanda, 1992); Cough, Nervous disorders (Kapur & Nanda, 1992); Asthma, Liver troubles, Diarrhoea, Fever (Janmeda *et al.,* 2006); Cuts (Kapur & Nanda, 1992; Sharma, 2003).

Whole plant - Cardiac disorders (Gohil & Qadri, 1992).

Root - Kidney trouble (Gohil & Qadri, 1992).

ORIGANUM MAJORANA L. (LAMIACEAE)

Syn.: *Majorana hotensis* Moench.

Regional & Other Names : *English* – Sweet majoram; *Hindi*-Marua; *Sanskrit* – Marubaka.

Distribution: Cultivated throughout India.

Taxonomic Description: Root stock perennial, shortly creeping; the annual stems erect, more or less hairy. Leaves ovate-lanceolate, slightly toothed. Flowers Purple.

Medicinal Parts: Whole plant, Leaves, Seeds & Essential oil.

Chemical Constituents: Leaves - Majoranin, Luteolin, Apigenin, Diosmetin.

Medicinal Properties:

मरूबकः कटुष्णश्च दीपनस्तिक्ततीक्ष्णकः ।
हृदयः पित्त करो रूच्यो रूक्षो लघुसुगन्धिकः ।। (नि. र.)

Marūbakaḥ katuṣṇaṣca dīpanaṣtiktatīkṣṇakaḥ.
Hṛdyaḥ piṭtakaro rūcyo rūkso laghusugaṇdhikaḥ. (Ni. Ra.)

Antipyretic, Astringent, Carminative, Diaphoretic, Emmenagogue, Expectorant, Liver tonic, Stimulant.

Medicinal Uses: Burns, General weakness, Headache, Liver compliant, Scorpion sting, Sprains, Stomach trouble, Tooth trouble, Worms.

ORYZA SATIVA L. (POACEAE)

Regional & Other Names : *Bengali* – Dhaan; *English* – Paddy; *Gujarati* - Choka; *Hindi* – Chawal; *Kannada* – Akki; *Malayalam* - Ari; *Marathi* – Tandula; *Sanskrit* –Syali; *Tamil* – Arishi; *Telugu* – Urlu.

Distribution: Much cultivated in wet fields throughout India.

Taxonomic Description: Tall, annual grass, panicles lax, at length drooping, spikelets persistent. Pale-brown at maturity.

Medicinal Parts: Roots, Grains.

Chemical Constituents: Grain contains Protein, Carbohydrate, Fibre; Minerals; Essential Amino acids, Free amino acids: Alamine, Tyrosine, Proline, Glutamic acid, Aspartic acid, Cystien, Vitamins.

Medicinal Properties:

व्रीहयः कथिता पाके मधुरा वीर्यतो हिताः ।
अल्पाभिष्यन्दिनो बद्धवर्चस्काः षष्टिकैः समाः ।।
कृष्णव्रीहीर्वरस्तेषां तस्मादल्पगुणाः परे ।। (भा. प्र.)

Vrīhayaḥ kathitā pāke madhurā vīryato hitāḥ
Aḷpābhiṣyaṇdino baddhavaṛcaṣkāḥ saṣtikaīḥ samāḥ
Kṛṣṇavṛīhīṛvaṛstesāṃ taṣmādaḷpaguṇāḥ parae. (Bhā. Pṛa.)

Grains - Acrid, Aphrodisiac, Diuretic, Sweet, Tonic.

Medicinal Uses: Grains - Biliousness.

Ethno- Medicinal Uses: Flower- Antifertility (Sharma, 1981).

OUGEINIA OOJEINENSIS (Roxb.) Hochr. (FABACEAE)

Syn.: *O. dalbergioides* Benth.

Regional & Other Names : *Bengali* – Tinis; *English* – Sandan; *Gujarati* - Tanach; *Hindi* – Sand Sandan an; *Kannada* – Karimutale; *Malayalam* - Totukara; *Marathi* – Timas; *Sanskrit* – Tinisah; *Tamil* – Narivengai; *Telugu* – Tellamotuku.

Distribution: Throughout India.

Taxonomic Description: A medium sized deciduous trees, with liliac or White flowers and flat pods.

Medicinal Parts: Bark.

Chemical Constituents: Bark contains-Triterpenoids as lupeol, Betulin, β-amyrin & Hydroxylupeol.

Medicinal Properties:

तिनिशस्तु कषायोष्णः कफरक्तातिसारजित् ।
ग्राहको दाहजननो वातामयहरः परः ।। (रा. नि.)

Tiniṣaṣtu kasāyoṣṇaḥ kapharaktātisārajiṭ
Gṛāhako dāhajanano vātāmayaharaḥ paraḥ. (Rā. Ni.)

Bark - Acrid, Astringent, Anthelmintic, Anti-inflammatory, Constipating, Cooling, Depurative, Febrifuge, Stimulant, Styptic, Sudorific.

Medicinal Uses: Bark - Anaemia, Burning sensation, Diarrhoea, Dysentery, Fever, Inflammation, Leprosy, Leucoderma, General debility, Gonorrhoea, Ulcers.

Ethno- Medicinal Uses: Stem bark – Asthma, Bronchitis (Joshi *et al.,* 1980); Dysentery (Maheshwari *et al.,* 1986; Sinha *et al.,* 2006); Urinary infections, leucoderma (Chhetri, 2005); Diarrhoea (Sinha *et al.,* 2006).

Leaf - Bronchitis (Joshi *et al.,* 1980).

OXYSTELMA ESCULENTUM R. Br. (ASCLEPIADACEAE)

Syn.: *O. secamone* (L.) Karst.; *Periploca secamone* L.; *Periploca esculenta* L.f.

Regional & Other Names : *Bengali* – Dudhialata; *Gujarati* - Jaldudhi; *Hindi* – Dudlata; *Kannada* –Dugdhike; *Marathi*–Dugdhani; *Sanskrit* – Dughika; *Tamil* –Usippalai; *Telugu* – Dudipalla.

Distribution: Throughout the plains & lower hills of India.

Taxonomic Description: A glabrous twiner. Leaves thin, narrowly lanceolate; flowers Whitish with Purple. Follicles glabrous, ovoid-lanceolate, pointed.

Medicinal Parts: Whole plant, Roots, Milky juice, Fruits.

Medicinal Properties:

दुग्धिका स्वादुपर्णी स्यात्क्षीरा विक्षीरिणी तथा ।
दुग्धिकोष्णा गुरू रूक्षा वातलागर्भकारिणी ।। (भा. प्र.)

Dugdhikā śvādupaṛṇī ṣyāṭksirā viksīriṇī tathā
Dugthikoṣṇā gurū ruksā vātalā gaṛbhkāriṇī. (Bhā. Pṛa.)

Whole plant - Anthelmintic, Aphrodisiac, Bitter, Diuretic, Dry, Hot, Indigestible, Lactagogue, Laxative, Pungent.

Fruits – Anthelmintic, Bitter, Expectorant, Tonic.

Medicinal Uses: Whole plant - Bronchitis, Leaucoderma, Sore throat, Ulcers.

Roots - Jaundice.

Fruits - Cough, Gleet, Gonorrhoea, Leucodema, Muscular pain.

Ethno- Medicinal Uses: Root - Jaundice ((Baruah *et al.,* 1987; Shah, 1984; Shah *et al.,* 1983); Liver complaint (Shah *et al.,* 1983).

Root - Jaundice (Sinha *et al.,* 2006).

PANICUM SUMATRENSE Roth. ex Roem. et Schult. (POACEAE)

Syn.: *Panicum miliare* Lam.

Regional & Other Names : *Bengali* – Gundli; *English* – Little millet; *Gujarati* – Cheno; *Hindi* – Savan; *Kannada* – Save; *Malayalam* - Cama; *Marathi* –Sava; *Sanskrit* – Syamakah; *Tamil* – Camai; *Telugu* – Korralu.

Distribution: Cultivated throughout India.

Taxonomic Description: An annual grass.

Medicinal Parts: Grains.

Chemical Constituents: Grains - Albuminoids, Starch, Oil, Fibre.

Medicinal Properties:

श्यामाको मधुरः स्निग्धः कषायो लघु शीतलः ।
वातकृत् कफपित्तघ्नः संग्राहि विषदोषनुत् ।। (रा. नि.)

Śyāmāko madhuraḥ ṣnigdhaḥ kasāyo laghu śītalaḥ
Vatakṛt kaphapiṭtaghnaḥ saṃgṛāhi visadosanuṭ. (Rā. Ni.)

Grains - Alexeteric, Astringent, Constipating, Cooling, Digestive, Emollient, Sweet.

Medicinal Uses: Grains - Burning sensation, Diarrhoea, Dysentery, Dyspepsia, Gonorrhoea, Ulcers, Wounds.

Ethno- Medicinal Uses: Px-Nervine stimulant & Tonic (Kumar & Yadav, 2006).

Grains - Diarrhoea, Dysentery, Wounds, Ulcers, Gonorrhoea (Basak, 2006).

PASPALUM SCROBICULATUM L. (POACEAE)

Syn.: *P. commersonii* Lam.

Regional & Other Names : *Bengali* – Kodoadhan; *English* – Punctured paspalum; *Gujarati* - Harik; *Hindi* – Kodo; *Kannada* – Haraka; *Malayalam* - Varaku; *Marathi* – Kadra; *Sanskrit* – Hudala; *Tamil* – Allu; *Telugu* –Arikalu.

Distribution: Throughout India.

Taxonomic Description: Erect or ascending, perennial grass, often rooting at lower nodes. Leaves tapering to a fine point. Racemes digitate, with 2-ranked spikelets. Caryopsis subglobose.

Medicinal Parts: Grains, Whole plant.

Chemical Constituents: Grains contains - Amino acids: Arginine, Histidine, Isoleucine, Leucine, Lysine, Methionine, Phenylalanine, Threonine, Tryptophan, Valine, N-lysine, Hentriacontanol, Hentriacontanone, Sitosterol, campesterol.

Medicinal Properties:

कोद्रवो वातलो ग्राही हिमा पित्तकफापहः ।
उद्दालस्तु भवेदुष्णो ग्राही वातकरो भृशम् ।। (भा. प्र.)

Koḍravo vātalo grāhi himā piṭtakaphāpahaḥ
Uḍdālaṣtu bhaveduṣṇo grāhi vātakāro bhṛśaṃ. (Bhā. Pṛa.)

Grains - Alexteric, Astringent, Bitter, Constipating, Cooling, Diuretic, Sedative, Styptic, Sweet, Tonic.

Medicinal Uses: Grains - Diarrhoea, Faltulence, Haemorrhages, Inflammation, Strangury, Ulcers.

Ethno- Medicinal Uses: Grains - Cold, cough (Maheshwari & Singh, 1987); Ulcer, Inflammation, Diseases of liver (Kumar & Yadav, 2006).

PENTAPETES PHOENICEA L. (STERCULIACEAE)

Regional & Other Names : *Bengali* – Kat-lata; *English* – Noon plant; *Gujarati* - Duporio; *Hindi* – Dopahariya; *Kannada* – Bandure; *Malayalam* - Uccamalari; *Marathi* –Tambdi-dupari; *Sanskrit* – Bandhukah; *Tamil* – Uccittiliakam; *Telugu* – Makinacettu.

Distribution: In North-Western India, Bengal & Gujarat, also cultivated in gardens.

Taxonomic Description: Annual herbaceous plants, Pink or Red flowers and capsule fruits.

Medicinal Parts: Roots, Fruits.

Medicinal Properties:

बन्धूकं श्लेष्मळं ग्राहि वातपित्तहरं लघु ।। (कै. नि.)

Baṇdhukam ́slesmalam grahi vatapittaharam laghu (Kai. Ni.)

Roots - Astringent, Constipating, Febrifuge, Mildly Thermogenic.

Fruits - Mucilaginous.

Medicinal Uses: Roots - Burning sensation, Diarrhoea, Fever.

Fruits - Fever, Gastropathy.

Ethno- Medicinal Uses: Bronchitis, Cough, Dysentery, Fever (Jain & Tarafder, 1970).

PHASEOLUS MUNGO (L.) Mant. (FABACEAE)

Syn.: *Vigna mungo* (L.) Hepper.

Regional & Other Names : *Bengali* – Mashkalai; *English* – Black gram; *Gujarati* -Adad; *Hindi* – Urid; *Kannada* – Uddinabee; *Malayalam* - Uzhunnuparippu; *Marathi* – Uddachi; *Sanskrit* – Masha; *Tamil* – Ulunnu; *Telugu* – Minapapappu.

Distribution: Commonly cultivated throughout India.

Taxonomic Description: Trailing, hispid herbs, with 3-foliolate leaves & yellow flowers. Pods linear, turgid.

Medicinal Parts: Whole plant, Roots.

Chemical Constituents: The plant contains Albuminoids, Starch and Oil.

Seed contains Saponin I, II & III. It contains Genistein, □-Hydroxygenistein, Dalbergioidin etc.

Medicinal Properties:

माषो गुरूः स्वादुपाकः स्निग्धो रूच्योनिलापहः ।
ऊष्णः सन्तर्पणो वल्यः शुक्रलो वृंदणः परः ।।
माषपर्णी हिमा तिक्ता स्निग्धा शुक्रबलासकृत् ।।
मधुरा ग्राहिणी शोथवातपित्तज्वरास्त्रजित् ।। (भा. प्र.)

Māso gurūḥ śvādupākaḥ ṣnigdho rūcyonilāpahaḥ
Uṣnaḥ saṇtaṛpaṇo baḷyaḥ śukṛalo bṛṇhaṇaḥ paraḥ
Māsapaṛṇī himā tiktā śnigdhā śukṛabalāsakṛt
Madhurā gṛāhiṇī śothavātapiṭtajvarāṣṛajiṭ. (Bhā. Pṛa.)

Root - Narcotic.

Seed - Aphrodisiac, Appetiser, Diuretic, Laxative, Sweet, Tonic.

Medicinal Uses: Seeds - Asthma, Cough, Dysentery, Diarrhoea, Dyspepsia, Paralysis, Piles, Rheumatism.

Ethno- Medicinal Uses: Seeds - Spermatorrhoea (Singh, 2003); Gout (Jadeja, *et al.*, 2006).

PHOENIX PUSILLA Gaertn. (ARECACEAE)

Syn.: P. farinifera Roxb.

Regional and Other Names: *English* – Small wild date-palm; *Hindi* – Palavat; *Kannada* – Indu; *Malayalam* – Cittintal; *Sanskrit* – Parusakah; *Tamil* – Cittintu; *Telugu* – Cinna.

Distribution: In the dry forests of Kerala, Karnataka, Tamil Nadu.

Taxonomic description: A shrubby stoloniferous palm with a very short stem enveloped in the sheath of the leaves; leaves pinnate; flowers in much branched spadix enclosed in spathes, Male flowers white & female greenish, not on the same plant; fruits Berries.

Medicinal parts: Fruits.

Taste : Sweet, Sour.

Medicinal properties:

परूशकं परूशकमल्पास्थि च परापरम् ।। (भा प्र.)

"Purusakaṃ parusakamaḷpāṣthi ca parāparaṃ." (Bhā. Pṛa.)

Fruits - Aphrodisiac, carminative, cooling, laxative, sour, sweet.

Medicinal uses: Fruits - Burning sensation, fever, general debility.

PHYLLANTHUS RETICULATUS Poir. (EUPHORBIACEAE)

Syn.: Kirganelia reticulata (Poir.) Ball.

Regional & Other Names : *Bengali* – Pon-kushi; *Hindi* – Pancoli; *Kannada* – Pulaveri; *Malayalam* - Niruri; *Sanskrit* – Krsnakamboji; *Tamil* – Karinelli; *Telugu* – Pulaguvva.

Distribution: Throughout India.

Taxonomic Description: A large, diffuse or straggling, glabrous shrub. Branchlets drooping. Leaves oblong, elliptic or rotundate, glabrous. Male flowers Purplish, in fascicles. Female flowers solitary. Berries globose, dark Purple.

Medicinal Parts: Whole plant.

Chemical Constituents: Leaves & Bark contains - Tannic acid, Gum & a Crystalline Principle.

Medicinal Properties:

कषाया शीतला शोफघ्नी च मधुरा कृष्णकाम्बोजी ।। (स्व.)

Kasāyā śītalā śophaghnī ca madhurā kṛṣṇakāṃbojī. (Ṣva.)

Alterant, Astringent, Constipating, Cooling, Diuretic, Stomachic, Sweet.

Medicinal Uses: Burning sensation, Burns, Diarrhoea, Gastropathy, Obesity, Skin eruptions, Sores, Strangury.

Ethno- Medicinal Uses: Root - Cough (Mishra & Sahu, 1984).

Leaf - Diarrhoea, Diuretic (Mishra & Sahu, 1984); Gum bleeding (Mishra & Sahu, 1984; Sudhakar & Rolla, 1985; Das *et al.,* 1983); Veneral diseases (Ravikumar & Sankar, 2003).

Fruits - Jaundice (Sinha *et al.,* 2006).

PIPER BRACHYSTACHYUM Wall. (PIPERACEAE)

Syn.: *Chavica sphaerostachya* Miq. *Piper mullesua* D. Don.

Regional & Other Names : *English* – Wild pepper; *Hindi* – Cab; *Kannada* – Carya; *Malayalam* - Kattumulaku; *Sanskrit* – Cavyah; *Tamil* – Kattumilagu; *Telugu* – Caikama.

Distribution: In the Western ghats, Nilgiris, Anamalais and hills of South Kerala.

Taxonomic Description: Woody climber; leaves elliptic-ovate or lanceolate, 3-5 ribbed, shortly petioled; fruiting spikes globose; fruits small ovoid or globose, 1-seedes berries.

Medicinal Parts - Roots.

Chemical Constituents: Plant contain - 2 unsaturated amides - brachystamides A & B.

Medicinal Properties:

चव्यं च कटुकोष्णं स्यात् जन्तुह्रत् दीपनं परम् ।
कफोद्रेकहरं वातकोपनं शमनं भवेत् ।। (ध. नि.)

Cavyaṃ ca katukoṣṇaṃ ṣyāṭ jantuhṛt dīpanaṃ paraṃ
Kaphodṛekaharaṃ vātakopanaṃ śamanaṃ bhaveṭ. (Dha. Ni.)

Roots - Acrid, Alexeteric, Anthelmintic, Appetising, Aromatic, Bitter, Carminative, Digestive, Expectorant. Stomachic, Thermogenic.

Medicinal Uses: Roots – Anorexia, Asthma, Bronchitis, Dysentery, Dyspepsia.

Ethno- Medicinal Uses: Fruit & Root – Cold (Das *et al.,* 1983); Cough (Das *et al.,* 1983; Karkongor & Joseph, 1981).

PLUMBAGO INDICA L. (PLUMBAGINACEAE)

Syn.: *P. rosea* L.

Regional & Other Names : *Bengali* – Lal chitra; *Gujarati* -Lal chitrak; *Hindi-* Chitra; *Kannada* –Kempacitramulam; *Malayalam* -Chivappukoduveli; *Marathi* –Lal chitrak; *Sanskrit* –Chitraka; *Tamil* – Cithiramulam; *Telugu* – Errachitramulam.

Distribution: Throughout India.

Taxonomic Description: A perennial erect, subscandent shrubs, with Bright Red flowers and capsules fruits.

Medicinal Parts: Root, Root bark, Milky jice.

Chemical Constituents: Root contains Plumbagin, β-sitosterol, Arachidic alcohol, Toxic agents like, m-dinitrobenzene, Alphanapthylamine, □-Naphthoguinone.

Medicinal Properties:

स्थूलकायकरो रूच्यो कुष्ठध्नो रक्तचित्रकः ।
रसेनियामको लोहे बेधकश्च रसायनः ।। (रा. नि.)

Ṣthulakāyakarō rūcyo kuṣthaghno raktaciṭrakaḥ
Rase niyāmako lohe vedhakaṣca rasāyanaḥ. (Rā. Ni.)

Root - Abortifacient, Acrid, Alterant, Anthelmintic, Anti-inflammatory, Antiperiodic, Astringent, Carminative, Constipating, Digestive, Diuretic, Narcotic, Rejuvenating, Stimulant, Sudorific, Thermogenic.

Medicinal Uses: Root - Amenorrhoea, Anaemia, Bronchitis, Colic, Cough, Diabetes, Dyspepsia, Elephantiasis, Fever, Flatulence, Haemorrhoids, Helminthiasis, Inflammation, Leprosy, Leucoderma, Oedema, Piles, Ringworm, Scabies, Skin diseases, Worms.

Ethno- Medicinal Uses: Latex - Headache (Mudgal & Pal, 1980).

Root – Abortifacient (Tarafder, 1983; Bhalla *et al.*, 1982; Saxena & Dutta, 1975; Prasad *et al.*, 2003); Piles, Dysentery (Jadeja *et al.*, 2006); Rheumatism (Mudgal & Pal, 1980).

PODOPHYLLUM EMODI Wall. ex Hook. f. & Th. (BERBERIDACEAE)

Syn.: *P. hexandrum* Royle.

Regional & Other Names : *Bengali* – Papra; *English* – Indian podophyllum; *Gujarati* -Venivel; *Hindi* – Bakrachimaka; *Marathi* – Patvel; *Sanskrit* – Hansapadi.

Distribution: Himalayas, Shimla, Nepal.

Taxonomic Description: A glabrous perennial herb growing in damp shady places, with a creeping knotty rhizome bearing thick fibrous roots & small Red fruits.

Medicinal Parts: Whole plant.

Chemical Constituents: It contains a Resin with Essential oil, Podophyllotoxin, Picropodophyllin and Quercetin. Dried rhizomes yield Podophyllotoxin-B-D-Glucoside. Roots Rhizomes yield □-lignan, a Glycoside, Astragalin (Kaempferol-3-Glucoside) Isolated.

Medicinal Properties:

हंसपादी गुरूः शीता हन्ति रक्तविषव्रणम् ।
विसर्पदाहातिसारलूताग्निरोहिनीः ।। (भा. प्र.)

Haṇsapādī guruḥ śītā haṇti raktavisaṿraṇaṃ
Visaṛpadāhātisāralūtābhutagnirohinīḥ. (Bhā. Pṛa.)

Alterative, Bitter, Emetic, Purgative, Tonic.

Medicinal Uses: Bilious complaints, Constipation.

Ethno- Medicinal Use: Rhizome - Cancer (Rawat & Pangtey, 1987; Gantait *et al.,* 2006); Cuts, wounds (Uniyal & Chauhan, 1973; Gantait *et al.,* 2006); Diarrhoea (Srivastava *et al.,* 1981; Gupta *et al.,* 1981; Gantait *et al.,* 2006); Fever (Nautiyal, 1981; Gantait *et al.,* 2006); Gastric ulcers (Uniyal & Chauhan, 1973); Hepatic diseases (Gaur *et al.,* 1983; Srivastava *et al.,* 1981; Gupta *et al.,* 1981); Cough, Gastric complaint, Skin diseases (Gantait *et al.,* 2006).

Fruits - Cough (Koelz, 1979); Skin lesions (Gohil & Qadri, 1992); Blood diarrhoea (Kala & Manjrekar, 2003).

Resin - Anticancer, Anti-tubercular glands (Arya & Agarwal, 2006).

Root - Skin problems, to regulate menstrual cycle & Burning sensation of urine in women (Ballabh & Chaurasia, 2006).

POLYGONATUM CIRRHIFOLIUM (Wall.) Royle (LILIACEAE)

Syn.: *P. sibiricum* Baker

Regional & Other Names : *English* – Meda; *Hindi* – Meda; *Kannada* – Medha; *Malayalam* - Meda; *Sanskrit* – Meda; *Tamil* – Meda; *Telugu* – Meda.

Distribution: In the Himalayas, from 1200 to 4200 m.

Taxonomic Description: A herb with thick root-stock, upper part of stem creeping, leafy and grooved. Leaves in whorls of 4, sessile, linear; flowers purplish tinged with white and green; fruits berries.

Medicinal Parts: Root stock.

Chemical Constituents: Rhizome contains - 2 steroidal saponins: Sibiricoside A (PS-I) and Sibiricoside B (PS-IV) in addition to neoprazerigenin A-3-O-β-lycotetraoside (PS-III) and its Methyl proto-type congeher (PS-II).

Medicinal Properties:

मेदा तु मधुरा शीता पित्तदाहार्त्तिकासनुत् ।
राजयक्ष्मज्वरहरा वातदोषकरी च सा ।। (रा. नि.)

Medā tu madhurā śītā piṭtadāhāṛttikāsanuṭ
Rājayakṣmajvaraharā vātadosakarī ca sā. (Rā. Ni.)

Root stock - Aphrodisiac, Cooling, Depurative, Emollient, Expectorant, Galactagogue, Sweet, Vulnerary.

Medicinal Uses: Root stock - Agalactia, Burning sensation, Bronchitis, Cough, Fever, General debility, Skin diseases, Tuberculosis, Ulcers, Wounds.

Ethno- Medicinal Uses: Rhizome - Chest pain, Dysentery, Jaundice (Janmeda *et al.*, 2006).

POLYGONATUM VERTICILLATUM (L.) All. (LILIACEAE)

Regional & Other Names : *English* – Mahameda; *Hindi* – Mahameda; *Kannada* – Mahamedha; *Malayalam* - Mahameda; *Sanskrit* – Mahameda; *Tamil* – Mahameda; *Telugu* – Mahameda.

Distribution: In the Himalayas, at a height of 1,800 to 3,900 m.

Taxonomic Description: Erect glabrous herb with thick root stock, upper part of the stem leafy & grooved; leaves whorled, Sessile, linear; flower Pinkish white; fruits berries.

Medicinal Parts: Root stock.

Chemical Constituents: Plant contains a glucoside of digitalis group.

Medicinal Properties:

महामेदा हिमा स्वादुः कफशुक्लविवर्धनी ।
हन्ति दाहास्रपित्तानि क्षयवातज्वरैः सह ।। (ध. नि.)

Mahāmedā himā ṣvāduḥ kaphaśuklavivaṛdhanī
Haṇti dāhāsṛapiṭtāni ksayavātajvaraīḥ saha. (Dha. Ni.)

Root stock - Aphrodisiac, Appetising, Cooling, Diuretic, Emollient, Galactagogue, Sweet, Tonic.

Medicinal Uses: Root stock - Anorexia, Burning sensation, Fever, General debility, Seminal weakness, Strangury.

Ethno- Medicinal Uses: Root - Nervine tonic, Tonic (Uniyal, 1968; Gupta, 1981); Aphrodisiac (Gupta, 1981); Tuberculosis & Asthma (Kapur & Nanda, 1992); Diabetes, Sexual impotency (Arya & Agarwal, 2006); To restore body strength, Kidney problems (Ballabh & Chaurasia, 2006).

PRUNUS ARMENIACA L. (ROSACEAE)

Regional & Other Names : *English* – Common apricot; *Hindi* – Chuari; *Sanskrit* – Aluka.

Distribution: Cultivated in North India.

Taxonomic Description: A moderate sized tree.

Medicinal Parts: Seed, Oil.

Chemical Constituents: Seed contains Fatty oil, Vitamins, Enzyme, Amygdalin.

Leaves contain Quercetin, Cyanidin, Kaempferol, Caffeic acid, P-coumaric acid.

Medicinal Properties:

गुरूष्णास्निग्ध मधुर सोरूमान बलप्रदः ।।

Gurūṣṇaṣnigdha madhura sorūmāna balapradaḥ.

Fruit - Antidiarrhoeal, Antipyretic, Emetic, Sweet.

Seeds – Anthelmintic, Nutritive, Tonic.

Medicinal Uses: Seed – Deafness, Diseases of liver, Earache, Piles.

Ethno- Medicinal Uses: Kernel oil - Rheumatic pains (Sharma & Rana, 2003).

PSEUDARTHRIA VISCIDA (L.) Wight & Arn. (FABACEAE)

Regional & Other Names : *Gujarati* – Chapakno-velo; *Hindi* – Salparni; *Kannada* – ; *Malayalam* - Muvvila; *Marathi* –Salparni; *Sanskrit* – Salparni; *Tamil* – Nirmalli; *Telugu* – Muyyakuponna.

Distribution: Throughout South India.

Taxonomic Description: Perennial semi-erect diffuse undershrubs, with Pink flowers & Pods fruits.

Medicinal Parts : Roots.

Chemical Constituents: Root - Leucopelargonidin.

Medicinal Properties:

वातघ्नी बृंहणी वृष्या क्षयकासज्वरापहा ।
रसायनी स्वादुरसा शालपर्ण्यतिसारजित् ।। (म. नि.)

Vātaghnī bṛṃhaṇī vṛṣyā ksayakāsajvarāpahā
Rasāyanī ṣvādurasā śālapaṛṇyatisārajịt. (Ma. Ni.)

Roots - Anthlemintic, Anti-Inflammatory, Aphrodisiac, Astringent, Bitter, Constipating, Digestive, Diuretic, Emollient, Febrifuge, Sweet, Thermogenic.

Medicinal Uses: Roots - Asthma, Bronchitis, Cough, Diabetes, Diarrhoea, Dyspepsia, Fever, General debility, Gout, Helminthiasis, Inflammation, Strangury, Tuberculosis.

Ethno- Medicinal Uses: Leaf - Women for lactation (Binu *et al.,* 2003).

Root - Piles (Thomas & Britto, 2003).

PTEROSPERMUM CANESCENS Roxb. (STERCULIACEAE)

Syn.: *P. suberifolium* (Lin.) Lam.

Regional & Other Names : *Bengali* – Muchkand; *Hindi* – Mucukund; *Kannada* – Toddagiringa; *Malayalam* - Mucukundam; *Marathi* – Muchkand; *Sanskrit* – Mucukundah; *Tamil* – Vennanku; *Telugu* – Lolagu.

Distribution: Throughout South India.

Taxonomic Description: Medium-Sized trees, with fragrant White flowers & capsules fruits.

Medicinal Parts: Flowers.

Chemical Constituents: β-sitosterol and a mixture of Fatty acids containing myristic, Palmitic, Stearic, Arachidic, Oleic & Linoleic acids from flowers.

Medicinal Properties:

मुचुकुन्दः कटुस्तिक्तः कफकासविनाशश्च कण्ठदोषहरः ।
त्वग्दोषशोफशमनो व्रणपामाविनाशनश्चैव ।। (रा. नि.)

Mucukundaḥ katuṣtiktaḥ kaphakāsavināśaṣca kanthadōsaharaḥ
Tvagdōṣaśōphaśamanō vranapāmāvināśanascaiva. (Ra. Ni.)

Flowers - Acrid, Alexeteric, Alexeteric.Anodyne, Astringent, Bitter, Depurative, Mucilagenous.

Medicinal Uses: Flowers - Bronchitis, Cough, Inflammation, Sores, Leprosy, Skin diseases.

RANUNCULUS SCELERATUS L. (RANUNCULACEAE)

Regional & Other Names : *English* – Blister buttercup; *Hindi*-Jal dhania; *Sanskrit* – Kandira.

Distribution: Northern India, Mount Abu, Warms Valleys of Himalaya, Bengal.

Taxonomic Description: Annual herbs, with White flowers & achene fruits.

Medicinal Parts: Whole plant.

Chemical Constituents: Seed contain a Fatty oil.

Medicinal Properties:

काण्डीरः कटुतिक्तोष्णः सरोदृष्टव्रणार्त्तिजित् ।
लतागुल्मोदरप्लीहशूलमन्दाग्निनाशनः ।। (ध. नि.)

Kandiraḥ Katutiktosnaḥ sarodustavrinārttijịt
Lalāgulmodarplīhashulmaṇdāgnināsanaḥ (Dha. Ni.)

Whole plant - Acrid, Diuretic, Stimulant.

Medicinal Uses: Whole plant - Asthma, Dysuria, Pneumonia, Rheumatism, Sciatica.

Seeds - Kidney troubles , Stomachic, Tonic.

Ethno- Medicinal Uses: Leaf - Eczema (Saxena & Vyas, 1981).

RHAPHIDOPHORA LACINIATA (Burm. f.) Merr. (ARACEAE)

Syn.: *R. pertusa* (Roxb.) Schott.

Regional & Other Names : *Bengali* – Gajpipal; *Gujarati* - Motipipal; *Kannada* – Dodda thippali; *Malayalam* - Elittatiyan; *Marathi* – Ganeshkanda; *Sanskrit* – Sphotyabhujangam; *Tamil* – Anaittippili; *Telugu* – Enugan alleru.

Distribution: Throughout India.

Taxonomic Description: A semi-epiphytic climbing shrubs, with Greenish-Yellow flowers & Berries fruits.

Medicinal Parts: Stem.

Medicinal Properties:

स्फोट्याभुजङ्.गं कटुतिक्तं कषायं दीपनपाचनम् ।। (स्व.)

Ṣphoṭyābhujaṇgaṃ katutiktaṃ kasāyaṃ dīpanpācnaṃ. (Ṣva.)

Stem - Acrid, Astringent, Bitter, Carminative, Digestive.

Medicinal Uses: Stem – Bronchiopathy, Colonalgia, Ulcers.

RHINACANTHUS NASUTA (L.) Kurz. (ACANTHACEAE)

Syn.: *R. communis* Nees.

Regional & Other Names : *Bengali* – Juipana; *Hindi* – Palakjuhi; *Kannada* – Nagamulli; *Malayalam* - Nagamulla; *Marathi* – Gajakarni; *Sanskrit* – Gajakarni; *Tamil* – Nagamalli; *Telugu* – Nagamulli.

Distribution: Throughout India.

Taxonomic Description: An undershrubs, with White flowers and velvety-pubescent capsules fruits.

Medicinal Parts: Whole plant.

Chemical Constituents: Root - Rhinocanthin.

Medicinal Properties:

गजकर्णी स्वादुपाका तिक्ता चोष्णा कफापहा ।
शीतज्वरं च वातं च नाशयेदिति कीर्तिता ।। (नि. र.)

Gajakarṇī ṣvādupākā tiktā coṣṇā kaphāpahā
Śītajvaraṃ ca vātaṃ ca nāśayediti kīṛtitā. (Ni. Ra.)

Acrid, Anthelminitc, Anticancerous, Antiparasitic, Antiseptic, Aphrodisiac, Bitter, Depurative, Thermogenic.

Medicinal Uses: Cancer, Eczema, Helminthiasis, Inflammation, Leprosy, Obesity, Ringworm, Scurvy, Septic Ulcers.

Ethno-Medicinal Uses: Root - Eczema, Ringworm (Janardhanan, 1963).

RHUS PARVIFLORA Roxb. (ANACARDIACEAE)

Regional & Other Names : *Hindi* – Tung; *Sanskrit* – Tintidika.

Distribution: North-Western Himalaya from the Sutlej to Nepal 2,500-5,000 ft., Panchamarhi hills in the Central provinces, Rampa hills in the Godavari district.

Taxonomic Description: A shrub, with Green flowers & drupe fruits.

Medicinal Parts: Fruits.

Medicinal Properties:

वातापहं तिन्तिडीकमामं पित्तबलासकृत् ।
ग्राहयुष्णं दीपनं रूच्यं सपक्वं कफवातनुत् ।। (सु. सू.)

Vatāpahaṃ tiṇtidikmāṃ pittabalāskruṭ
Grāhyusnaṃ dīpanaṃ rucyaṃ sapakvaṃ kaphavātanuṭ. (Su. Su.)

Medicinal Uses: Dental disorders, Conjuctivitis, Oedema.

Ethno- Medicinal Uses: Root Bark - Sunstroke (Jain,1984).

ROTULA AQUATICA Lour. (BORAGINACEAE)

Syn.: *Rhabdia lycioides* Sensu C.B. Clarke non Mart.

Regional & Other Names : *Kannada* – Pasanabheda; *Malayalam* - Kalurvanci; *Marathi* – Sherni; *Sanskrit* – Pasanabhedah; *Tamil* – Seppunerinji; *Telugu* – Pasanabheda.

Distribution: Throughout India.

Taxonomic Description: Small shrubs, with Pink or Reddish flowers and drupes fruits.

Medicinal Parts: Roots.

Chemical Constituents: Bauerenol, α - & β - amyrin.

Medicinal Properties:

अश्मभेदो हिमस्तिक्तः शर्कराशिश्नशूलजित् ।
हृद्रोगप्लीहगुल्मार्शो वस्तिशुद्धिकरः परः ।। (ध. नि.)

Aśmabhedo himastiktaḥ śaṛkarāśiśnaśūlajiṭ
Hṛdrogaplīhaguḷmāṛśo vaṣtiśuddhikaraḥ paraḥ. (Dha. Ni.)

Root - Astringent, Bitter, Cooling, Diuretic, Laxative.

Medicinal Uses: Root - Haemorrhoids, Diabetes & Venereal diseases, Renal and Vesical Calculi,

Ethno-Medicinal Uses: Whole plant - Piles, Veneral diseases (Sinha *et al.,* 2006).

RUNGIA REPENS Nees. (ACANTHACEAE)

Regional & Other Names : *Gujarati* - Khatsalio; *Hindi* –Kharmor; *Kannada* – Kodagasale; *Marathi*–Ghatipitpapada; *Sanskrit* – Parpatha; *Tamil* – Kodagasaleh.

Distribution: Throughout India.

Taxonomic Description: A spreading, procumbent herb, rooting from lower nodes. Leaves lanceolate to oblong-elliptic. Flowers Violet or Pinkish, in erect, terminal, Pretty spikes. Capsules hairy.

Medicinal Parts: Whole plant.

Chemical Constituents: Flowers, Luteolin-7-glycoside and Delphinidin, Besides Luteolin and another Flavone-Chrysoeriol.

Medicinal Properties:

पर्पटःशीतलस्तिक्तः संग्राही वातकोपनः ।
लघुः पाके च कटुको हरेत्पित कफत्पित कफज्वरान् ।। (नि. र.)

Paṛpatah śītalastiktaḥ saṇgrāhī vātakopanaḥ
Laghuḥ pāke ca katuko harẹtpitta kaphạtpitta kaphajvarāṇ. (Ni. Ri.)

Plant - Diuretic, Vermifuge.

Medicinal Uses: Plant - Cough, Fever, Snake bite.

Ethno- Medicinal Uses: Whole plant - Cough, Cold, Fever (Singh & Pandey, 1998).

SACCHARUM ARUNDINACEUM Retz. (POACEAE)

Syn.: *Erianthus arundinaceus* (Retz.) Jesw. ex Heyne

Regional & Other Names : *English* – Devil sugarcane; *Hindi* – Ramsar; *Kannada* – Kadukabba; *Malayalam* - Ama; *Sanskrit* – Munjah; *Tamil* – Munji; *Telugu* – Adaviceruku.

Distribution: Throughout India.

Taxonomic Description: Found along River banks and Water logged areas. Stems glabrous, Smooth, Internodes, Very long; Leaves deep Green;

Inflorescence large, Pink, White or Silvery; Spikelets in pair, one sessile and the other pedicelled.

Medicinal Parts: Root, stem.

Medicinal Properties:

मुञ्जस्तु मधुरः शीतः कफपित्तजदोषजित् ।
ग्रहरक्षासु दीक्षासु पावनो भूतनाशनः ।। (रा. नि.)

Maṇjaṣtu madhuraḥ śītaḥ kaphapiṭtajadosajiṭ
Gṛaharaksāsu dīksāsu pāvano bhūtanāśanaḥ. (Rā. Ni.)

Roots - Demulcent, Diuretic.

Stem - Aphorodisiac, Astringent, Depurative, Refrigerant, Sweet.

Medicinal Uses: Stem - Burning sensation, Dipsia, Erysipelas, Haemopathy, Uropathy.

Ethno- Medicinal Uses: Root - Post delivery cases (Dhiman, 2003); Urinary system troubles, Kidney stones (Chhetri, 2005).

SACCHARUM MUNJA Roxb. (POACEAE)

Syn.: *Erianthus munja* Jesw.

Regional & Other Names : Bengali – Muncha; Hindi – Munj; Marathi – Mole; Sanskrit – Sara; Telugu – Gundra.

Distribution: North India in the Punjab & Upper Gangetic plain.

Taxonomic Description: Perennial grass, pale violet or Green coloured-spikelete arranged in panicle inflorescence, fruits rectangular.

Medicinal Parts: Roots.

Medicinal Properties:

मुञ्जद्वयं तु मधुरं तुवरं शिशिरं तथा ।
दाहतृष्णाविसर्पास्रमूत्रकृच्छ्राक्षिरोगजित् ।। (भा. प्र.)

Munchajdvayaṃ tu madhuraṃ tuvaraṃ śiśiraṃ tathā
Dāhatrusnāvisarpāsramutrakrucchākshirogajiṭ. (Bha. Pra.)

Stem - Acrid, Aphrodisiac, Cooling, Sweet.

Medicinal Uses: Stem - Blood troubles, Burning sensation, Erysipelas, Eye diseases, Urinary complaints.

SACCHARUM SPONTANEUM L. (POACEAE)

Regional & Other Names : *English* – Thatch grass; *Hindi* – Kus; *Malayalam* - Kusa; *Sanskrit* – Kasah; *Tamil* – Pekkarimpu; *Telugu* – Kakiceruku.

Distribution: Throughout India.

Taxonomic Description: A very variable, tall, perennial, densely tufted grass. Leaves very long, narrow-linear, acuminate, coriaceous. Flowering panicles conical or lanceolate to oblong; peduncle softly silky just below the panicle. Spikelets paired, one pedicellate, the other sessile, 1-flowered, hermaphrodite.

Medicinal Parts: Roots.

Medicinal Properties:

कासः स्यान्मधुरस्तिक्तः स्वादुपाको हिमः सरः ।
मूत्रकृच्छ्राश्मदाहास्रक्षतपित्ताक्षिरोगजित् ।। (भा. प्र.)

Kāsaḥ ṣyāṇmadhuraṣtiktaḥ ṣvādupāko himaḥ saraḥ
Mūṭrakṛcchrāśmadāhāsṛaksatapiṭṭāksirogajiṭ. (Bhā. Pṛa.)

Roots - Aphrodisiac, Astringent, Diuretic, Laxative, Refrigerant, Sweet, Tonic.

Medicinal Uses: Roots - Burning sensation, Dysentery, Dysentery, Dyspepsia, General debility, Haemorrhoids, Menorrhagia, Renal & Vesical Calculi, Strangury.

Ethno- Medicinal Uses: Px - Asthma, Burns, Cholera, Cough, Urinary Complaint (Jain & Tarafder, 1970); Urinary disorders, Calculus, Uterine bleeding (Kapur & Nanda, 1992); Vesicle calculi, Diseases of blood (Kumar & Yadav, 2006).

Whole plant - Diseases of blood, Biliousness (Mukhopadhyay, 1992).

Root - Toothache (Satapathy, 1992).

Young shoot - Impotency (Barua *et al.,* 2003).

SALACIA RETICULATA Wight. (HIPPOCRATEACEAE)

Regional & Other Names : *Kannada* – Ekanayakam; *Malayalam* - Ekanayakam; *Sanskrit* – Vairi; *Tamil* – Ponkoranti; *Telugu* – Anukuducettu.

Distribution: Throughout India.

Taxonomic Description: Large woody climbers, with pale Yellow bark & Yellowish Green flowers.

Medicinal Parts: Roots.

Medicinal Properties:

वैरी शीतः पित्तरक्तव्रणघ्नो मेहजित् परम् ।। (ह्र. प्रि.)

Vaīri śītaḥ piṭtaraktavṛaṇaghno mehajịt paraṃ. (Hṛ. Pṛi.)

Roots - Acrid, Anodyne, Anti-inflammatory, Bitter, Depurative, Inflammation, Liver tonic, Stomachic, Thermogenic, Urinary astringent, Vulnerary.

Medicinal Uses: Roots - Amenorrhoea, Colic, Diabetes, Dysmenorrhoea, Dyspepsia, Flatulence, Haemorrhoids, Inflammation, Leprosy, Leucorrhoea, Skin diseases, Spermatorrhoea, Ulcers, Wounds.

Ethnomedicinal Uses: Roots - Skin diseases (Bhandary & Chandrashekar, 2002).

SALVADORA PERSICA L. Var. *WIGHTIANA* [Planch ex Thw.] Verdc. (SALVADORACEAE)

Syn.: *S. indica* Wight.

Regional & Other Names : *Bengali* – Jhal; *English* –Tooth brush tree; *Gujarati* - Kharijal; *Hindi* –Jhak; *Kannada* –Goni mara; *Marathi* –Miraj; *Sanskrit* - Angahava; *Tamil* –Kalawa; *Telugu* – Ghunia.

Distribution: Drier parts of India.

Taxonomic Description: A small or medium sized, crooked or bent. Leaves variable in shape ovate, elliptic-ovate or ovate lanceolate, pale Green, glossy, fleshy. Flowers Greenish-Yellow, in axillary & terminal panicles. Drupes globose, fleshy, dark Red at Maturity.

Medicinal Parts: Bark, Leaves, Shoots, Fruits.

Chemical Constituents: Root - Alkaloid-Trimethylamine, β-Sitosterol, Glucotropaeolin, Monoclinic Sulphur.

Stem- β-Sitosterol, Octacosanol, Salvadoside, Salvadoraside, Siringin.

Leaves - Salvadoricine.

Medicinal Properties:

पीलुः श्लेष्मसमीरघ्नं पित्तलं भेदि गुल्मनुत् ।
स्वादुतिक्तं च यत्पीलु तन्नात्युष्णं त्रिदोषहृत् ।। (भा. प्र.)

Pīluḥ śleṣmasamīraghnaṃ piṭtalaṃ bhedi gụlmanuṭ
Ṣvādutiktaṃ ca yạtpīlu tạnnāṭyuṣṇaṃ ṭridosahṛṭ. (Bhā. Pṛa.)

Leaves - Anthelmintic, Antiscorbutic, Astringent, Bitter, Diuretic, Expectorant, Tonic.

Fruits - Acrid, Alexeteric, Aphrodisiac, Bitter, Digestive, Emollient, Purgative, Stomachic, Sweet, Thermogenic.

Medicinal Uses: Leaves - Asthma, Bronchitis, Constipation, Cough, Haemorrhoids, Scurvy, Strangury, Verminosis.

Fruits - Constipation, Gastropathy, Seminal weakness.

Ethno- Medicinal Uses: Leaf - Asthma, Cough (Shah, 1982).

Fruit - Constipation (Shah, 1982).

Root - Ringworm (Mishra *et al.,* 1976).

Bark - Pain in Back, Joints, limbs (Shah, 1982; Chelladurai, 1983).

SAMADERA INDICA Gaertn. (SIMAROUBACEAE)

Syn.: *Quassia indica* (Gaertn.) Nooteboom.

Regional & Other Names : *English* – Niepa bark tree; *Hindi* – Lokhanti; *Kannada* – Nipa; *Malayalam* - Karinnotta; *Marathi* – Lokhandi; *Sanskrit* – Gucchakaranjah; *Tamil* – Nibam; *Telugu* – Lokanti.

Distribution: Throughout South India.

Taxonomic Description: Small trees, with pale Yellow bark & Pinkish-Yellow flowers.

Medicinal Parts: Bark, Wood, Seeds.

Chemical Constituents: Bark - Triterpene as Taraxerone; Sterols as Stigmastanone, Stigmasterol, quassinoids as Samaderines A, B, C, E, Samaderol, Lupenone, Indaquassin A, B, C, D, E, F; Dihydrosamaderine B, Brucein D, Soulameolide, Cedronin.

Medicinal Properties:

करञ्जः कटुतिक्तोष्णो विषवातार्त्तिकृन्तनः ।
कण्डूविचर्चिकाकुष्ठस्पर्शत्वग्दोषनाशनः ।। (रा. नि.)

Karaṇjaḥ katutiktoṣṇo visavātāṛttikṛṇtanaḥ
Kaṇdūvicaṛcikākuṣthaṣpaṛśaṭvagdosanāśanaḥ. (Rā. Ni.)

Bark / Wood - Stomachic, Febrifuge, Tonic.

Seed oil - Acrid, Astringent, Febrifuge, Depurative, Emetic, Purgative, Thermogenic.

Medicinal Uses: Bark/Wood-Colic, Dysmenorrhoea, Dyspepsia, Flatulence, General debility.

Leaves - Erysipelas, Pruritus.

Seed oil - Bilious fever, Constipation, Leprosy, Pruritus, Scabies, Skin diseases.

SAPINDUS LAURIFOLIA Vahl. (SAPINDACEAE)

Syn.: *S. trifoliata* Sensu *Hiern non* L.

Regional & Other Names :*Bengali* – Bara ritha; *English* – Soapnut tree; *Gujarati* -Aritha; *Hindi* – Ritta; *Kannada* – Kugatemara; *Malayalam* - Urunci; *Marathi* –Rithe; *Sanskrit* – Phenilah; *Tamil* – Puvankottai; *Telugu* – Kunkutikaya.

Distribution: Throughout India.

Taxonomic Description: Medium sized deciduous trees, with White flowers & Drupe fruits.

Medicinal Parts: Root, Bark, Fruits.

Chemical Constituents: Acyclic Sequiterpene Oligoglycoside: Trifolioside II, Saponin A, Sapindoside B, Aglycones: Hederagenin, Oleanolic acid, Sapindic acid.

Seeds - Fatty oil: Glycerides Palmitin, Stearic, Eicosanin, Lignocerin, Olein, Cyanolipids Containing Oleic acids & Arachidic acid; Phenolic acids: Protocatechic acid, Cis- P-coumaric acid, P-hydroxy benzoic acid, Cinnamic acid.

Medicinal Properties:

रीठा करञ्जः तिक्तोष्णः कटुस्निग्धश्च वातजित् ।
कफघ्नः कुष्ठकण्डूतिविषविस्फोटनाशनः ।। (रा. नि.)

Rīthā karaṇjaḥ tiktoṣṇaḥ katuṣnigdhaśca vātajiṭ
Kaphaghnaḥ kuṣthaṇdūtiviṣavisphotanāśanaḥ. (Rā. Ni.)

Roots & Bark – Demulcent, Expectorant.

Fruits - Abortifaceint, Acrid, Anthlemintic, Astringent, Bitter, Emetic, Expectorant, Thermogenic, Tonic.

Medicinal Uses: Roots – Epilepsy, Hysteria.

Fruits - Asthma, Cholera, Diarrhoea, Lumbago, Verminosis.

Ethno- Medicinal Uses: Fruits - Mumps (Kshirsagar & Singh, 2003); Emetic, Purgative (Nadanakunjidam, 2003); Dysentery (Jadeja *et al.,* 2006).

Fruit juice - Asthma, Hysteria, Epilepsy (Deokule, 2006).

Leaves, Stem Bark - Dandruff (Ghosh, 2006).

SARCOSTEMMA ACIDUM (Voigt) Voigt (ASCLEPIADACEAE)

Syn.: *S. brevistigma* Wight & Arn.

Regional & Other Names : *Bengali* – Somlata; *English* – Moon plant; *Gujarati* - Somlata; *Hindi* – Somlata; *Kannada* – Somalata; *Malayalam* – Somalata; *Marathi* – Somvel; *Sanskrit* – Somavalli; *Tamil* – Somam; *Telugu* – Kondapala.

Distribution: Bihar, Bengal & South India, in dry rooky places upto 1,350 m.

Taxonomic Description: Perennial leafless, jointed shrubs, with milky latex and fragrant White flowers.

Medicinal Parts: Whole plant.

Chemical Constituents: Lupeol acetate, □-amyrin acetate, β-amyrin-acetate, β-sitosterol, Malic acid, Succinic acid, Sucrose, Tannins, Sarcobiose, Brevobiose, Tigmobiose.

Medicinal Properties:

सोमवल्ली त्रिदोषघ्नी कटुस्तिक्ता रसायनी ।। (भा. प्र.)

Somavaḷlī tṛidosaghnī katuṣtiktā rasāyanī. (Bhā. Pṛa.)

Whole plant - Acrid, Alterant, Bitter, Cooling, Emetic, Narcotic, Rejuvenating.

Medicinal Uses: Whole plant - Dipsia, General debility, Hydrophobia, Viral infection.

Ethno- Medicinal Uses: Root - Antifertility (Sharma, 1981).

Latex - Boils (Ramachandran & Nair, 1981).

Plant paste - Bone fracture (Saxena, 1986).

SARCOSTIGMA KLEINII Wight & Arn. (ICACINACEAE)

Regional & Other Names : *Kannada* – Puvanna; *Malayalam* - Ota; *Sanskrit* – Ingudi; *Tamil* – Otal.

Distribution: At low elevations of the Western ghats in Kerala.

Taxonomic Description: Large climbing shrubs, with small Yellow unisexual flowers & drupes fruits.

Medicinal Parts: Bark, Leaves, Oil.

Chemical Constituents: Seeds yield a fatty oil.

Medicinal Properties:

इङ्.गुदी कफरक्तामग्रन्थिघ्नी स्याद् व्रणेहिता ।
ऐङ्.गुदं स्वादुतिक्तं च स्निग्धोष्णं श्लेष्मवातजित् ।। (सो. नि.)

Iṇgudī kapharaktāmagṛaṇthighnī ṣyāḍ vṛaṇehitā
Aiṇgudaṃ ṣvādutiktaṃ ca ̦snigdhoṣṇaṃ śleṣmavātajịt. (So. Ni.)

Bark / Leaves - Acrid, Anaphrodisiac, Anthlemintic, Bitter, Carmnative, Digestive, Diuretic, Depurative, Stomachic, Thermogenic, Vulnerary..

Oil - Anaphrodisiac, Anthelmintic, Bitter, Depurative, Vulnerary.

Medicinal Uses: Bark / Leaves - Anorexia, Epilepsy, Flatulence, Gastropathy, Helminthiasis, Hysteria, Leprosy, Skin diseases, Strangury.

Oil - Foul ulcers, Helminthiasis, Leprosy, Skin diseases.

SASSUREA LAPPA C.B. Clarke. (ASTERACEAE)

Regional & Other Names : *English* – Kuth; *Hindi* – Kuth; *Kannada* – Kostha; *Malayalam* - Kottam; *Sanskrit* – Kusthah; *Tamil* – Kostam; *Telugu* – Kostamu.

Distribution: In Kashmir, at an altitude between 2,500-4,000 m.

Taxonomic Description: Erect perennial plants, with Bluish Purple flowers & achenes fruits.

Medicinal Parts: Roots.

Chemical Constituents: Roots - Resin, Essential oil, Alkaloids, Inulin, Fixed oil, Tannins, Sugars.

Medicinal Properties-

कुष्ठं तिक्तं कटु स्वादु लघुष्णं शुक्ललं जयेत् ।
वातास्रविषवीसर्पकुष्ठकासकफानिलान् ।। (कै. नि.)

Kuṣthaṃ tiktaṃ katu.svādu laghuṣṇaṃ śuklalaṃ jayeṭ
Vātāsṛavisavīsaṛpakuṣthakāsakaphānilāṇ. (Kaī. Ni.)

Roots - Acrid, Alterant, Anodyne, Aphrodisiac, Aromatic, Bitter, Carminative, Deodorant, Depurative, Diaphoretic, Digestive, Diuretic, Expectorant, Febrifuge, Stimulant, Stomachic, Sweet, Thermogenic, Tonic.

Medicinal Uses: Roots - Asthma, Bronchitis, Cephalagia, Cholera, Colic, Cough, Dyspepsia, Erysipelas, Flatulence, Gout, Leprosy, Leucoderma, Strangury.

Ethno- Medicinal Uses: Root - Stomach pain, Blood purifier, Infection, Dryness, Boils (Arya *et. al.*, 1999); Rheumatic pain (Ara & Naqshi, 1992); To restore body strength, Kidney problems (Ballabh & Chaurasia, 2006); Dysentery, Ulcer, Stomachache (Kala & Manjrekar, 2003).

SCHLEICHERA OLEOSA (Lour.) Oken. (SAPINDACEAE)

Syn.: *S. trijuga* Willd.

Regional & Other Names : *Bengali* – Kusum; *English* – Lac tree; *Gujarati* - Kosano; *Hindi* – Kusum; *Kannada* – Sagade; *Malayalam* – Puvam; *Marathi* – Gosam; *Sanskrit* – Mukulakah; *Tamil* – Puvattipuvam; *Telugu* – Posuku.

Distribution: Throughout India.

Taxonomic Description: A handsome tree. Leaves paripinnate. Flowers Yellowish-Green. Fruits pointed, Smooth or echinate.

Medicinal Parts: Bark, Oil.

Chemical Constituents: Seed - Phenols, Polyene Pigments.

Seed oil - Cyanolipids, Oleic acid, Gadoleic acid, Stearic acid, Arachidic acid, Behenic acid, Palmitic acid, Palmitoleic acid.

Leaves - Stigmasterol, β-sitosterol, Campesterol, Cholesterol, Fructose, Glucose, Tartaric acid, Oxalic acid.

Medicinal Properties:

केशाम्रदंतिनीस्नेहाः कषायकटुतिक्तकाः ।
कुष्ठव्रणहराः श्लेष्मकृमिवातापहाः सराः ।। (सो. नि.)

Kośāmṛadaṃtinīṣnehāḥ kasāyakatutiktakāḥ
Kuṣthavṛaṇaharāḥ śleṣmakṛmivatāpahāḥ sarāḥ. (So. Ni.)

Bark – Antipyretic, Astringent.

Seed oil - Anthelmintic, Bitter, Purgative, Sour, Stomachic, Sweet, Tonic.

Medicinal Uses: Bark - Inflammation, Malaria, Pruritus, Ulcers.

Seed oil - Burns, Cephalalgia, Ulcers.

Ethno- Medicinal Uses: Bark – Gout, Ithching, Paralysis (Gupta, 1981); Rheumatism (Gupta, 1981; Maheshwari *et al.,* 1986; Tribedi *et al.,* 1982); Sprain (Jain & De, 1966).

Seed - Rheumatism (Shah *et al.,* 1981; Roy & Chaturvedi, 1986); Skin diseases (Shah *et al.,* 1981; Jain 1965; Pal & Srivastava, 1976).

Seed oil - Skin diseases (Nadanakunjidam, 2003); Burns (Jain 1965; Roy & Chaturvedi, 1986); Wounds (Roy & Chaturvedi, 1986; Maheshwari *et al.*, 1981).

Stem - Skin diseases, Itching, Eruption (Sasidharan & Augustine, 2006).

SCHREBERA SWIETENIOIDES Roxb. (OLEACEAE)

Regional & Other Names : *Bengali* – Ghantaparul; *English* – Weaver's beam tree; *Gujarati* - Mokho; *Hindi* – Moka; *Kannada* – Bula; *Malayalam* – Malamplasu; *Marathi* – Kokadi; *Sanskrit* – Muskakah; *Tamil* – Mogalingam; *Telugu* – Magalinga.

Distribution: Throughout India.

Taxonomic Description: Moderate sized deciduous tree, with Yellowish-Brown flowers & capsule fruits.

Medicinal Parts: Root, Bark, Leaves, Fruits.

Chemical Constituents: Leaves - Acetic acid, Formic acid, Methyl alcohol, β-sitosterol.

Medicinal Properties:

मोक्षकः कटुकस्तिक्तो ग्राहयुष्णः कफवातहृत् ।
विषमेदोगुल्मकण्डूवस्तिरुक्कृमिशुक्लहृत् ।। (भा. प्र.)

Moksakaḥ katukaṣtikto grāhyuṣṇaḥ kaphavātahṛt
Visamedogulmakaṇḍūvaṣtirukkrmiśuklahṛt. (Bhā. Pṛa.)

Root / Bark / Leaves - Bitter, Acrid, Appetising, Digestive, Thermogenic, Stomachic, Depurative, Constipating, Urinary astringent, Anthelmintic.

Fruits - Digestive, Purgative, Stomachic.

Medicinal Uses: Root / Bark / Leaves - Anaemia, Boils, Burns.

Colic, Diarrhoea, Dyspepsia, Flatulence, Leprosy, Skin diseases.

Fruits - Anaemia, Anorexia, Colic, Diabetes, Flatulence.

Ethno- Medicinal Uses: Bark - Cough (Jain 1965); Mental depression (Maheshwari *et al.*, 1986); Stomachahce. (Prasad *et al.*, 2003).

Fruits - Hydrocoel (Jain & De, 1966; Mudgal & Pal, 1980; Pal & Srivastava, 1976).

Root - Leprosy (Sudhakar & Rolla, 1985).

SCINDAPSUS OFFICINALIS (Roxb.) Schott. (ARACEAE)

Syn.: *Pothos officinalis* Roxb.

Regional & Other Names : *Bengali* – Gajpipalusum; *Gujarati* - Gajpiper; *Hindi* – Gajapipal; *Kannada* – Gajahippali; *Malayalam* – Attitti-ppali; *Marathi* – Hattipipli; *Sanskrit*– Gajapippali; *Tamil* – Attittippili; *Telugu* – Gajapippallu.

Distribution: Throughout India.

Taxonomic Description: Large epiphytic climbers, with Spadix inflorescence & Berries fruits.

Medicinal Parts: Dried mature inflorescence.

Chemical Constituents: Inflorescence - Scindapsinidine-Scindapsinidine-B, Glucose, Cyclopropenoid, Fatty acids, Free sugars, Fructose, Rhamnose, Xylose, Disaccharide, Trisaccharide.

Medicinal Properties:

हस्तिकृष्णा कटुः पाके वीर्योष्णा दीपनी कटुः ।
वातश्लेष्मकृमिश्वासकण्ठरोगातिसारजित् ।। (कै. नि.)

Haṣtikṛṣṇā katuḥ pāke vīryoṣṇā dīpanī katuḥ
Vātaśleṣmakṛmiṣvāsakaṇtharogātisārajiṭ. (Kaī. Ni.)

Dried inflorescence - Acrid, Anthelmintic, Aphrodisiac, Carminative, Diaphoretic, Expectorant, Stimulant, Sweet, Thermogenic.

Medicinal Uses: Dried inflorescence - Bronchitis, Cough, Diarrhoea, Helminthiasis.

Ethno- Medicinal Uses: Whole plant - Bone fracture (Chaudhuri *et al.*, 1982).

Root – Cough, Stomachache (Saxena *et al.*, 1981).

Leaf - Bone fracture (Singh, 2003).

Fruits - Roundworms (Panda & Das, 2003).

SECURINEGA LEUCOPYRUS (Willd.) Muell-Arg. (EUPHORBIACEAE)

Syn.: *Fluggea leucopyrus* Willd.

Regional & Other Names : *Gujarati* - Shinwi; *Hindi* – Aints; *Kannada* – Bilchuli; *Malayalam* – Mulpulanni; *Marathi* – Vorepuvan; *Sanskrit* – Bhuriphali; *Tamil* – Mulluppulatti; *Telugu* – Puli.

Distribution: In dry localities throughout India.

Taxonomic Description: A large, woody, much branched, diffuse shrub. Branches White; branchlets spine-like, Purplish when young. Leaves less than 2.5 X 1.5 cm., obovate, elliptic or obcordate, glaucous. Male flowers Greenish-Yellow, slender, pedicelled Berries globose, White.

Medicinal Parts: Whole plant.

Chemical Constituents: Leaves contain - Triterpenes, Diterpenes, Steroids.

Stem bark - Tannins, Friedelin.

Medicinal Properties:

पाण्डुफली तु मधुरा बल्या वृष्या च शीतला ।
मूत्राघातं पित्तरोगं मूत्रकृच्छ्रास्ररूग्जयेत् ।। (नि. र.)

Pāṇḍuphalī tu madhurā baḷyā vṛṣyā ca śītalā
Mūṭrāghātaṃ piṭtarogaṃ mūṭrakṛcchṛāsṛarūgjayeṭ. (Ni. Ra.)

Whole plant - Aphrodisiac, Cooling, Diuretic, Sweet, Tonic.

Medicinal Uses: Whole plant - Burning sensation, General debility, Seminal weakness, Strangury.

Ethno- Medicinal Uses: Leaves - Sores, Wounds, Boils (Bhatt *et al.*, 2003); Stomachache (Subramani & Goraya, 2003); Gout (Jadeja *et al.*, 2006c).

SEMECARPUS ANACARDIUM L. f. (ANACARDIACEAE)

Regional & Other Names : *Bengali* – Bhela; *English* – Marking nut tree; *Gujarati* - Bhilamu; *Hindi* – Bhilawa; *Kannada* –Goddugeru; *Malayalam* – Chera; *Marathi* – Bibha; *Sanskrit* – Bhallataka; *Tamil* – Shenkottei; *Telugu* – Bhallatamu.

Distribution: Throughout India.

Taxonomic Description: A small trees, with unisexual flowers and drupe fruits.

Medicinal Parts: Fruits.

Chemical Constituents: Fruits contain - Anacardiac acid, Cardol, Catechol, Anacardol, Fixed oil, Somearpol, Bhilawanol.

Medicinal Properties:

भल्लातः कटुतिक्तोष्णो मधुरः कृमिनाशनः ।
गुल्मार्शोग्रहणीकुष्ठान् हन्ति वातकफामयान् ।। (ध. नि.)

Bhal̤lātaḥ katutiktoṣṇo madhuraḥ kṛmināśanaḥ
Guḷmāṛśog̣rahaṇī kuṣthāṇ haṇti vātakaphāmayāṇ. (Dha. Ni.)

Fruits - Acrid, Anthelmintic, Anti-Inflammatory, Aphrodisiac, Astringent, Bitter, Depurative, Digestive, Emollient, Expectorant, Febrifuge, Purgative, Sudorific, Sweet, Thermogenic, Tonic.

Medicinal Uses: Fruits - Asthma, Caridac diseases, Colic, Constipation, Cough, Diabetes, Dyspepsia, Fever, Flatulence, General debility, Haemorrhoids, Helminthiasis, Hepatopathy, Sciatica, Scrofula, Tumours, Ulcers.

Ethno- Medicinal Uses: Bark - Permanent sterility in female (Bhogaonkar & Kadam, 2006); Wounds caused due to burning (Satapathy & Brahman, 2003); Gonorrhoea (Tarafder, 1984).

Fruits – Asthma, Cough (Sudhakar & Rolla, 1985); Cuts (Jain 1965); Dyspepsia, (Bhalla *et al.,* 1982); Epilepsy (Bhalla *et al.,* 1982; Sudhakar & Rolla, 1985; Roy & Chaturvedi, 1986); Nervous debility (Bhalla *et al.,* 1982; Roy & Chaturvedi, 1986); Rheumatism (Sudhakar & Rolla, 1985); Skin diseases (Bhalla *et al.,* 1982); Throat diseases (Shah *et al.,* 1983); Worms (Bhalla *et al.,* 1982); Abortifaceint, Asthma, Leprosy, Antifertility (Sinha *et al.,* 2006); Piles (Bhalla *et al.,* 1982; Nadanakunjidam, 2003; Sinha *et al.,* 2006).

Root – Abortifacient, Antifertility (Chaudhuri & Pal, 1975; Sinha *et al.,* 2006); Impotency (Maheshwari *et al.,* 1986; Prakash & Singh, 2006); Wounds (Prasad *et. al.,* 2003); Asthma, Leprosy, Piles, (Sinha *et al.,* 2006).

Seed - Abortifaceint (Chaudhuri & Pal, 1975); Rheumatism (Rai, 1985; Shah *et al.,* 1983); Skin diseases (Shah, 1984); Stomach pain (Sur *et al.,* 1992); Cuts, Wounds (Panda & Das, 2003).

Seed oil - Epilepsy (Chaudhuri & Pal, 1975); Rheumatism (Shah *et al.,* 1981); Ulcer (Chaudhuri & Pal, 1975); Skin diseases (Shah *et al.,* 1981; Chaudhuri & Pal, 1975); Swelling (Chaudhuri & Pal, 1975); Cough, Cold, Malaria (Sharma & Baishya, 1992).

Latex – Headache, Skin diseases (Singh & Maheshwari, 1983).

Fruit oil - Sprain (Sharma, *et al.,* 1984); Cuts, Wounds (Panda & Das, 2003).

Leaf - Boils (Panda & Das, 2003); Abortifaceint, Asthma, Piles, Antifertility (Sinha *et al.,* 2006); Leprosy (Roy & Chaturvedi, 1986; Sinha *et al.,* 2006); Eczema, Allergy (Prasad *et al.,* 2003).

SESAMUM INDICUM L. (PEDALIACEAE)

Regional & Other Names : *Bengali* – Til; *English* – Gingelly; *Gujarati* - Til; *Hindi*-Til; *Kannada* – Ellu; *Malayalam* – Ellu; *Marathi* – Til; *Sanskrit* – Tilah; *Tamil* – Aellu; *Telugu* – Nuvulu.

Distribution: Cultivated throughout India.

Taxonomic Description: An erect, hairy robust or tall herb. Lower leaves often compound, lobed or deeply devided, broad ovate; upper ones undevided, linear-oblong. Flowers Pink or rosy, bent, solitary; capusules erect oblong, valvular.

Medicinal Part : Roots, Leaves, Seeds, Oil.

Chemical Constituents: Leaves - A flavonoid as Pedailin & Pinoresinol.

Seeds - Sesamol, □-tocopherol, β-tocopherol, a lignan as Sesamin; a Nitrolactone as Sesamolin, Thiamine, Niacin, Globulin, Sesamolinol, Sesaminol, Pinoresinol, Sesamolinol.

Medicinal Properties:

स्निग्धोष्णो मधुरस्तिक्तः कषायः कटुकस्तिलः ।
त्वच्यः केश्यश्च बल्यश्च वातघ्नः कफपित्तकृत् ।। (च. सं.)

Ṣnigdhoṣṇo madhurạstiktaḥ kasāyaḥ katukaṣtilaḥ
Ṭvacyaḥ keśyaśca bạlyośca vātaghnaḥ kaphapiṭtakṛt. (Ca. Saṃ.)

Roots / Leaves - Emollient.

Seeds - Acrid, Aphordisiac, Astringent, Bitter, Digestive, Emollient, Laxative, Sweet, Thermogenic, Tonic.

Oil - Anthelmintic, Astringent, Bitter, Constipating Digestive, Emollient, Sweet, Thermogenic.

Medicinal Uses: Leaves - Cholera, Dermatopathy, Dysentery, Nephropathy, Ophthalmopathy.

Seeds - Burns, Dysentery, Haemorrhoids, Migrain, Obesity, Ulcers.

Oil - Burning sensation, Cephalalgia, Gonorrhoea, Obesity, Otalgia.

Ethno- Medicinal Uses: Seed - Abortifaceint (Tarafder, 1983); Boils (Tarafder, 1984); Lactation (Bhattacharjee *et al.,* 1980); Ulcers (Malhotra & Moorthy, 1973; Sharma *et al.,* 1979); Gout (Jadeja *et al.,* 2006); Allergy (Das *et al.,* 2006); Cough, Cold (Sharma *et al.,* 2006).

Seed oil - Urinary troubles (Shukla *et al.,* 1992); Piles (Deokule, 2006).

Leaf - Swelling, Abdominal pain (Sharma et al. 2006).

SETARIA ITALICA (L.) P. Beauv. (POACEAE)

Regional & Other Names : *Bengali* – Kakani; *English* – Deccan grass; *Gujarati* - Kang; *Hindi* – Bartiya; *Kannada* – Bilikorla; *Malayalam* – Tena; *Marathi* – Bavani; *Sanskrit* – Kanghuni; *Tamil* – Tenai; *Telugu* – Koralu.

Distribution: Cultivated throughout India.

Taxonomic Description: Herbaceous annual plants, with spikelets inflorescence & various coloured oblong or ellipsoid grains.

Medicinal Parts: Grains.

Chemical Constituents: Grains - Folic acid, Necotinic acid, Riboflavin, Thiamine, Albumin, Globulin, Glutenin, Prolamine, □-setarin, β-setarin, Alanine, Lutamic acid, Histidine, Isoleucine, Leucine, Lysine, Methionine, Proline, Phenylalanine, Threonine, Tryptophane, Valine, Phytic acid, Arachidic acid, Behemic acid, Linoleic acid, Oleic acid, β-alanine, Betaline.

Medicinal Properties:

कङ्.गुस्तु भग्नसन्धानवातकृद् बृंहणी गुरूः ।
रूक्षा श्लेष्महरातीव वाजिनां गुणकृद् भृशम् ।। (भा. प्र.)

Kaṇguṣtu bhagnasaṇdhānavātakṛd bṛṃhaṇī gurūḥ
Rūksā śleṣmaharātīva vājināṃ guṇakṛd bhṛśaṃ. (Bhā. Pṛa.)

Grains - Acrid, Aphrodisiac, Astringent, Cooling, Diuretic, Laxative, Sweet.

Medicinal Uses: Grains - Burning sensation, Fractures, Rheumatism.

Ethno- Medicinal Uses: Root - Dyspepsia (Chaudhuri *et al.,* 1982).

Grain - Rheumatic swelling (Shivakumar *et al.,* 2003); Burning sensation, Fractures, Rheumatism (Kumar & Yadav, 2006); Abortifacient (Tarafder, 1983; Abraham, 1981).

SIDA SPINOSA L. (MALVACEAE)

Regional & Other Names : *Bengali* – Bonmethi; *English* –Prickly Sida; *Gujarati* - Gangeti; *Hindi* – Gangern; *Kannada* – Kadumentyo; *Malayalam* – Mayirmanikkam; *Marathi* – Banbavari; *Sanskrit* – Bala; *Tamil* – Mairmanikkam; *Telugu* – Nagabala.

Distribution: Throughout India.

Taxonomic Description: A woody, much branched herb or undushrub. Stems and branches rough with spiny tubercles at the nodes below the petioles. Leaves ovate to ovate-lanceolate, crenate-serrate glabrous above. Flowers White. Ripe carples 5, toped by 2, erect, rough beaks.

Medicinal Parts: Whole plant.

Chemical Constituents: Alkaloid as β-Phenethylamine, Ephedrine, Vasicinol, Vasicinone, Vasicine, Choline, Hypaphorine, Methyl ether, Hypaphorine & Betaine.

Medicinal Properties:

महाबला तु हृद्रोगवातार्शः शोफनाशनी ।
शुक्लबृद्धिकरी बल्या विषमज्वरहारिणी ।। (रा. नि.)

Mahābalā tu hṛdṛogavātāṛśaḥśophanāśanī
Śuklavṛddhikarī baḷyā visamajvarahāriṇī. (Rā. Ni.)

Root - Antiperiodic, Aphrodisiac, Diaphoretic, Tonic.

Leaves - Emollient, Refrigerant.

Fruits - Astringent, Cooling.

Medicinal Uses: Root - Fevers.

Leaves – Gleet, Gonorrhoea.

Ethno- Medicinal Uses: Root - Antifertility (Lal & Lata, 1980).

SISYMBRIUM IRIO L. (BRASSICACEAE)

Regional & Other Names : *English* – London-Rocket; *Hindi* – Khubkalan; *Marathi* – Ran-tikhi; *Sanskrit* – Khubakala.

Distribution: Punjab, Rajputana, Peshawar, Baluchistan, Central Asia, Afganistan.

Taxonomic Description: Annual plant with Yellow flowers in Raceme inflorescence and long straight fruits.

Medicinal Parts: Seed.

Chemical Constituents: Seed yield a semi drying oil.

Medicinal Properties:

खाकसी सर्षपाभासा कटुष्णा पिच्छिला सरा ।
वातश्लेष्महरा बल्या स्वेदनी ज्वरकासनुत् ।। (द्र. गु. वि.)

Khākasī sarsapābhāsā katusnā picchitā sarā
Vataślesmaharā balyā svedanī jvarkāsnuṭ. (Dra. Gu. Vi.)

Seed - Expectorant, Febrifuge, Rubifacient.

Medicinal Uses: Seed - Asthma.

SOLANUM DULCAMARA L. (SOLANACEAE)

Regional & Other Names: *English*–Bitter sweet; *Sanskrit* – Kakmachi.

Distribution: From Kashmir to Garhwal and Sikkim to Baluchistan.

Taxonomic Description: Perennial woody climber, with Purple flowers & Berries fruits.

Medicinal Parts: Fruits, Stem, Twigs, Root.

Chemical Constituents: The aerial parts contain many alkaloids as Tomatidenol, Solasodine, Solodulicidine, □ &β-Solamarine, Solasonine, Soladulcine, Solamargine, Cholesterol, Cycloartanol, Campesterol, Stigmasterol, β-sitosterol.

Fruit contains Yamogenin, β-Solamargine, tigogenin, Diosgenin, Tomatidenol, Cycloartenol, Lanosterol, Cycloartanol, Lophenol, Cycloeucalenol, Obtusifoliol, Gramisterol, Citrostadienol, Cholesterol, Campesterol, Isofucosterol, Sitosterol, Stigmasterol.

Medicinal Properties:

काकजंघा च तिक्तोष्णा रक्तपित्तज्वरापहा ।
कृमिदोषहरा वर्ण्या विषदोषहरा मता ।। (धनवंतरी)

Kākjanghā ca tikṭtoṣṇā raktapiṭtajvarāpahā
Kṛmidosaharā vaṛṇyā visadosaharā matā. (Dhanvaṇtarī)

Fruits - Alterative, Antirheumatic, Antisyphilitic, Diuretic, Expectorant, Stimulant.

Twigs - Alterative, Antirheumatic, Diuretic, Resolvent.

Medicinal Uses: Fruits - Asthma, Bronchitis, Eczema, Gout, Liver disorders, Psoriasis, Rheumatism, Skin diseases, Syphilis, Warts.

Twigs - Asthma, Psoriasis, Rheumatism, Scrofula, Skin diseases, Ulcers, Urinary troubles, Whooping Cough.

Ethno- Medicinal Uses: Whole plant - Dropsy, Dysentery, Hemorrhages, Hay fever, Headache, Influenza, Measles (Rajan *et al.,* 2003).

SOLENA AMPLEXICAULIS (Lam.) Gandhi (CUCURBITACEAE)

Syn.: *S. heterophylla* Lour.; *Melothria heterophylla* (Lour.) Cogn.

Regional & Other Names: *Bengali* – Kudari; *Hindi* – Anatmul; *Kannada* – Bimpuli; *Malayalam* – Mudari; *Sanskrit* – Amlavetasah; *Tamil* – Pulivanji; *Telugu* – Tiddanda.

Distribution: Throughout India.

Taxonomic Description: Scandent perennial tendriller diocious plants, with Polymorphous leaves & small Yellow flowers.

Medicinal Parts: Tuberous Roots, Leaves, Seeds.

Chemical Constituents: Tubers - A steroid, Lignoceric acid, Tricosanoic acid, Behenic acid.

Medicinal Properties:

अम्लवेतसमत्यम्लं कषायोष्णं च वातजित् ।
कफार्शः श्रमगुल्मध्नमरोचकहरं परम् ।। (रा. नि.)

Aṃlavetasamatyaṃlaṃ kasāyoṣṇaṃ ca vātajiṭ
Kaphāṛśaḥsṛamaguḷmaghnamarōcakaharaṃ paraṃ. (Rā. Ni.)

Tuberous roots - Appetiser, Astringent, Carminative, Digestive, Diuretic, Expectorant, Purgative, Sour, Stimulant, Thermogenic.

Medicinal Uses: Tuberous roots - Anorexia, Asthma, Bronchitis, Cardiac disorders, Colic, Constipaiton, Dyspepsia, Flatulence, Gonorrhoea, Haemorrhoids, Hiccough, Renal and Vesical Calculi, Spermatorrhoea, Strangury, Vomiting.

Leaves - Allergic inflammation.

Ethno- Medicinal Uses: Root/ Tuber - Antifertility (Molla & Roy, 1985); Spermatorrhoea (Ved Prakash & Mehrotra, 1987); Tonic (Saxena, 1986); Impotency in man (Bhogaonkar & Kadam, 2006).

Leaf - Cuts (Jain & De, 1966); Diabetes (Ved Prakash & Mehrotra, 1987; Sinha *et al.,* 2006); Spermatorrhoea (Sudhakar & Rolla, 1985); Cuts, Asthma, Leprosy, Piles, Antifertility (Sinha *et al.,* 2006).

Fruits - Fever (Karkongor & Joseph, 1981); Diabetes, Cuts, Asthma, Leprosy, Piles, Antifertility (Sinha *et al.,* 2006).

Seed - Snake bite (Sen *et al.,* 1985)

Tuber - Impotency in man (Bhogaonkar & Kadam, 2006).

SONCHUS ARVENSIS auct. plur. non. L. (ASTERACEAE)

Regional & Other Names : *Bengali* – Banpalang; *English* – Hogweed; *Hindi* – Sadhi; *Telugu* – Nallatapata.

Distribution: Throughout India.

Taxonomic Description: Erect perennial herbs; basal leaves in a rosette, upper ones remote, entire or lyrate, pinnatilobed, semiamplexicaul base; heads pale yellow in corymbose panicles; achenes linear-oblong; pappus hairs white.

Medicinal Parts: Whole plant.

Chemical Constituents: Plant - Fatty acids: Stearic acid & Palmic acid; Ceryl alcohol, Choline, Invert sugar, Tartaric acid; Cobalt, Copper, Iron, Zinc.

Latex - Rubber, Mannitol, a bitter substance □-&-β -Lactucerols.

Medicinal Properties: Plant - Bitter, Diuretic.

Medicinal Uses: Plant - Chronic Fever.

Ethno- Medicinal Uses: Whole plant – Cholera, Dysentery Dysentery (Megoneitso, Rao, 1983); Cuts, Wounds (Baruah & Sarma, 1984); Fever, Jaundice (Sharma *et al.,* 1979).

Leaf - Cuts, Wounds (Megoneitso, Rao, 1983)

SORGHUM VULGARE Pers. (POACEAE)

Regional & Other Names : *Bengali* – Juar; *English* – Indian millet; *Hindi* – Juar; *Kannada* – Jola; *Malayalam* – Cholam; *Marathi* – Jwari; *Sanskrit*-Yavanala; *Tamil* – Cholam; *Telugu* – Jolnalu.

Distribution: Cultivated throughout India.

Taxonomic Description: A tall annual grass. Leaves broad linear with a prominent white midrib. Spikelets large, broad, hairy; pedicelled usually neuter, pedicels short.

Medicinal Parts: Seeds.

Chemical Constituents: Root - Hexacosanol, β -Sitosterol, Gulcose.

Shoot - Cyanogenetic glucoside.

Leaves - HCN.

Medicinal Properties:

यारनालो गुरू शीतो रूक्षो ग्राही रूचिप्रदः ।
वृष्यो मलस्तम्भकरः स्वादुः पित्तकफापहः ।।
रक्तरोगप्रदशमनो मुनिभिः पूर्वभीरितः ।। (नि. र.)

Yāranālo gurū śīto rūkso grāhī rūcipradaḥ
Vṛṣyo malạstaṃbhakaraḥ ṣvāduḥ piṭtakaphāpaḥ.
Raktarogapṛadaśhmano munibhịḥ pūṛvabhīritịḥ (Ni. Ra.)

Seed / Grains - Aphrodisiac, Constipating, Cooling, Demulcent, Diuretic,

Medicinal Uses: Grains - Biliousness diseases of blood, Piles, Tumours, Ulcers.

Ethno- Medicinal Uses: Grains - Grain- Baldness (Lal & Yadav, 1983).

SPONDIAS PINNATA (L.f.) Kurz. (ANACARDIACEAE)

Syn.: *S. mangifera* Willd.

Regional & Other Names : *Bengali* – Abhada; *English* – Wild mango; *Gujarati* - Abheda; *Hindi* – Junli-Am; *Kannada* – Ambate; *Malayalam* – Ampalam; *Marathi* – Ambada; *Sanskrit* – Amratakah; *Tamil* – Ampalam; *Telugu* – Amatas.

Distribution: Throughout India.

Taxonomic Description: A medium sized tree, Branching above. Leaflets 9-13, elliptic-oblong. Flowers Greenish-White, polygamous, in panicles. Drupes fleshy, Yellow.

Medicinal Parts: Root, Bark, Leaf, Fruits.

Chemical Constituents: ***Leaves*** - 24-Methylene-Cycloartanone, Stigma-4-en-3-one, β-Sitosterol, Lignoceric acid, β-Sitosterol, Glucoside.

Fruits - β-amyrin, Oleanolic acid, Polysaccharides, Aminoacids as Glycine, cystine, Serine, Alanine, Leucine.

Medicinal Properties:

आम्रातकफलं बृष्यं पित्तास्रकफवह्निकृत् ।
शीतं कषायं मधुरं किञ्चिन्मारूतकृत् ।। (ध. नि.)

Āmṛātakaphalaṃ vṛṣyaṃ piṭtāgrakapḥavahnikṛt
Śītaṃ kasāyaṃ madhuraṃ kiṇciṇmārūtakṛt. (Dha. Ni.)

Bark - Aromatic, Astringent, Refrigerant.

Leaf – Acidic, Aromatic, Astringent.

Unripe fruits - Aphrodisiac, Astringent, Sour, Thermogenic.

Ripe fruits – Constipating, Cooling, Emollient, Sweet, Tonic,

Medicinal Uses: Roots - Menstrual trouble.

Bark - Diarrhoea, Dysentery, Muscular rheumatism, Vomiting.

Ripe fruits - Diarrhoea, General debility.

Ethno- Medicinal Uses: Bark - Dysentery (Negi *et al.,* 1985; Sinha *et al.,* 2006); Rheumatism (Puri, 1983); Stomachache (Tarafder, 1983); Diarrhoea, Cholera, Earache, Itch (Sinha *et al.,* 2006).

Leaf – Earache (Negi *et al.,* 1985); Itching; Ringworm (Srivastava & Varma, 1981; Pandey & Rout, 2006); Dysentery, Diarrhoea, Cholera, Earache, Itch (Sinha *et al.,* 2006).

STERCULIA FOETIDA L. (STERCULIACEAE)

Regional & Other Names : *Bengali* – Jangli badam; *Hindi* – Jungalbadam; *Kannada* –Bhatala penari; *Malayalam* – Pinari; *Marathi* – Goldaru; *Tamil* –Arali; *Telugu* –Guttapubadamu.

Distribution: Throughout South India.

Taxonomic Description: Medium sized trees, with dull Orange, flowers and follicles fruits.

Medicinal Parts: Heartwood.

Chemical Constituents: Seeds yield a fatty oil.

Medicinal Properties:

रक्तदोषहरः पूतिदारूः व्रणकृमिकुष्ठनुत् ।। (स्व.)

"Raktadosaharaḥ pūtidārūḥ vṛaṇkṛmikuṣthanuṭ." (Ṣva.)

Heartwood - Depurative, Febrifuge, Vulnerary.

Medicinal Uses: Heartwood - Erysipelas, Fever, Helminthiasis, Leprosy, Skin diseases, Ulcers.

STEROSPERMUM CHELONOIDES DC. (BIGNONIACEAE)

Syn.: *S. personatum* (Hassk.) D. Chatterjee.

Regional & Other Names : *Bengali*–Ghunta; *English* – Trumpet flower; *Gujarati* - Padal; *Hindi* – Padhal; *Kannada* – Hude; *Malayalam* – Puppaatiri; *Marathi* – Padal; *Sanskrit* – Alivallabha; *Tamil*–Ambu; *Telugu* – Kalaagora.

Distribution: Throughout India, in Moist forests upto 1200 m. especially on Western ghats.

Taxonomic Description: Large deciduous trees, with Yellow or crimson flowers tinged into Red colour and cylindric capsules.

Medicinal Parts: Root, Leaves, Flowers, Fruits, Seeds.

Chemical Constituents: Leaves - Flavone as Stereolensin; Lapachol, Dinatin-7-Glucuronide, 6-O-Glucosylcutellarein, 6-O-Glucosylcutellarein.

Root bark - β-Sitosterol, n-Triacontanol, Specioside.

Heartwood - Lapachol, Dehydro-a-Lapachone, Dehydrotectol.

Medicinal Properties:

द्वितीया पाटला श्वेता निर्दिष्ठा काष्ठा पाटला ।
सा चैव श्वेतुकुम्भीका कुबेराक्षी फलेरूहा ।।
पाटलायाः गुणैस्तद्वत्किंचिन्मारूतकृद्भवेत् ।। (ध. नि.)

Dvitīyā pātalā ṣvetā nirdiṣthā kāṣthā pātalā
Sā caīva.svetakuṃbhīkā kuberāksī phalerūhā
Pātalāyāḥ gunaīṣtadvaṭkiṇciṇ mārūtakṛdbhaveṭ. (Dha. Ni.)

Root - Antiasthamatic, Anticancer, Anti-Inflammatory, Antitumour, Antiviral, Astringent, Cardiotonic, Cooling, Diuretic, Expectorant, Febrifuge, Tonic.

Leaves – Antimalarial, Antirheumatic, Antitumour.

Flowers - Aphrodisiac, Astringent, Cardiotonic, Sweet, Tonic.

Fruits - Antileprotic, Astringent, Cooling, Sweet.

Medicinal Uses: Roots - Anorexia, Asthma, Cough, Diarrhoea, Dyspepsia, Fever, General debility, Haemorrhoids, Hiccough, Piles.

Flowers- Bleeding, Cardiopathy, Burning sensation, General debility, Hiccough.

Fruits - Blood diseases, Hiccough, Leprosy, Strangury.

Ethno- Medicinal Uses: Root - Asthma (Malhotra & Moorthy, 1973; Sharma *et al.,* 1979); Cooling, Tonic (Basak, 2006).

Bark – Cholera, Dysentery, malaria, Stomachach (Rao & Jamir, 1982); Indigestion, liver complaint (Megoneitso, Rao, 1983); Cooling, Tonic (Basak, 2006).

Seed - Rheumatism (Tribedi *et al.,* 1985).

Flowers - Cooling, Tonic (Basak, 2006).

STEREOSPERMUM COLAIS (Buch.- Ham. ex Dillw.) Mabberley (BIGNONIACEAE)

Syn.: *S. personatum* (Hassk.) Chatterjee

Regional & Other Names : *Bengali* – Dharmar; *English* – Trumpet flower tree; *Gujarati* - Padhal; *Hindi* – Padaria; *Kannada* – Hude; *Malayalam* – Karintakara; *Sanskrit* – Aliballabha; *Tamil* – Ambu; *Telugu* – Godalipulsu.

Distribution: Throughout India.

Taxonomic Description: Large deciduous trees, with Yellow flower & Capsules fruits.

Medicinal Parts: Roots, Leaves, Fruits, Seeds.

Chemical Constituents: Leaves - A flavone as Stereolensin & Lapachol.

Root - Lapachol, β -Sitosterol.

Medicinal Properties:

फलपुष्पोपयोगेन हिक्काकासहरा कटुः ।
पाटली सकषाया तु तिक्तानिलकफापहा ।। (म. नि.)

Phalapuṣpopayogena hikkākāsaharā katuḥ
Pātalī sakasāyā tu tiktānilakaphāpahā. (Ma. Ni.)

Root - Acrid, Anodyne, Anti-Inflammatory, Aphrodisiac, Appetiser, Astringent, Bitter, Constipating, Diuretic, Expectorant, Febrifuge, Toinc.

Flowers - Aphrodisiac, Astringent, Cooling, Sweet, Tonic.

Fruits -Astringent, Cooling, Sweet.

Medicinal Uses: Roots - Asthma, Cough, Diarrhoea, Dyspepsia, Hiccough, Hyperacidity, Hyperdipsia, Renal and Vesical calculi, Strangury.

Leaves - Cooling, Malaria, Otalgia, Wounds.

Flower - Burning sensation, Hiccough, General debility.

Ethno- Medicinal Uses: Bark - Vomiting, Stomachache (Borthakur, 1976).

Root - Scorpion sting (Nadanakunjidam, 2003).

STREBLUS ASPER Lour. (MORACEAE)

Regional & Other Names : *Bengali* – Sheora; *English* – Siamese rough bush; *Hindi* – Daheya; *Kannada* – Mittimara; *Malayalam* – Parakam; *Marathi* – Paraya; *Sanskrit* – Akshadhara; *Tamil* – Piray; *Telugu* – Pakki.

Distribution - Throughout India.

Taxonomic Description: A small evergreen tree, sometimes a shrub, branching form base. Branches erect, leafy. Leaves elliptic or rhomboid or obovate, acute or acuminate, rough, margins subcrenate.

Medicinal Parts - Roots, Bark, Leaves, Latex.

Chemical Constituents: Root - β-Sitosterol.

Stem bark-□-Amyrin acetate, Lupeol, β-Sitosterol, Mansonin, Strebloside.

Leaves - β-Sitosterol, □-Amyrin, Luepol, n-Triacontane, Stigmasterol, Betulin, Oleanolic acid, Sioraside.

Medicinal Properties:

शाखोटवल्कलक्वाथं गोमूत्रेण युतं पिबेत् ।
श्लीपदानां विनाशाय मेदोदोषनिवृतये ।। (शा. नि.)

Śākhotavạlkalakvāthaṃ gomūtreṇayutaṃ pibeṭ
Ślīpadānāṃ vināśāya medodosanivṛttaye. (Śā. Ni.)

Root - Acrid, Anti-Inflammatory, Antiseptic, Astringent, Bitter, Constipating, Febrifuge, Thermogenic, Vulnerary.

Bark - Anti-Inflammatory, Constipating, Vulnerary.

Latex - Antiseptic, Astringent, Sedative, Styptic.

Medicinal Uses: Root -Boils, Bronchitis, Cough, Diarrhoea, Dysentery, Elephantiasis, Fever, Haemorrhages, Inflammation, Sinusitis, Syphilis, Ulcer.

Bark - Diarrhoea, Dysentery, Fever, Inflammation.

Leaf - Agalactia, Bubose, Swellings.

Latex - Adenitis, Haemorrhages, Neuralgia, Sore heels.

Ethno- Medicinal Uses: Leaf - Eye complaint (Jain & De, 1966).

Latex - Pneumonia (Jain 1965).

Bark – Stomachache, Urinary complaint (Jain & De, 1966).

STRYCHNOS POTATORUM L. f. (LOGANIACEAE)

Regional & Other Names : *Bengali* –Nirmali; *English* – Celaring Nut; *Hindi* – Nirmali; *Kannada* – Cilu; *Malayalam* – Tettamparal; *Sanskrit* – Katakah; *Tamil* – Tettamaram; *Telugu* – Cillacettu.

Distribution: West Bengal, Central & South India upto 1,200 m.

Taxonomic Description: Medium sized deciduous glabrous trees, with fragrant White flowers & berries fruits.

Medicinal Parts: Seeds.

Chemical Constituents: Seeds - Brucine, Diaboline, β-Sitosterol, Stigmasterol, Oleanolic acid and its 3B-acetate and a Saponin, Galactose and Mannose from Seeds.

Leaf & Bark - Isomotiol, Mixture of Sitosterol, Stigmasterol, Campsterol.

Medicinal Properties:

कतकस्य फलं नेत्र्यं जलनिर्मलताकरम् ।
वातश्लेष्महरं शीतं मधुरं तुवरं गुरू ।। (भा. प्र.)

Katakaṣya phalaṃ neṭryaṃ jalaniṛmalatākaraṃ
Vātaśleṣmaharaṃ śītaṃ madhuraṃ tuvaraṃ gurū. (Bhā. Pṛa.)

Seed - Alexeteric, Anthelmintic, Aphrodisiac, Appetiser, Astrignent, Bitter, Digestive, Dimulcent, Diuretic, Emetic, Refrigerant, Sweet, Tonic.

Medicinal Uses: Seed - Bronchitis, Burning sensation, Chronic diarrhoea, Conjunctivitis, Diabetes, Dipsia, Dysentery, Gastropathy, Gonorrhoea, Leucorrhoea, Strangury, Ulcers.

Root - Leprosy.

Ethno- Medicinal Uses: Root - Pain in body, Limbs, Muscles (Jain 1965).

Seed - Sedative (Mudgal & Pal, 1980); Urinary infection (Arinathan *et al.*, 2003).

Bark - Diuretic, Tonic, Asthma, Stomach diseases (Sinha *et al.*, 2006).

SYZYGIUM JAMBOS (L.) Alston. (MYRTACEAE)

Syn.: *Eugenia jambos* L.

Regional & Other Names : *Bengali* – Gulab-Jamb; *English* – Rose-apple; *Hindi* – Gulab-Jamun; *Kannada* – Pannerale; *Malayalam* – Campa; *Sanskrit* – Campeyah; *Tamil* – Campai; *Telugu* – Jambuneredu.

Distribution: Cultivated throughout India.

Taxonomic Description: A small tree, with very large handsome flowers and white globose fruits.

Medicinal Parts: Bark, Fruits.

Chemical Constituents: Leaves yield an essential oil.

Medicinal Properties:

चाम्पेयस्य त्वचा तिक्ता कषाया मधुरा तथा ।। (स्व.)

"Cāmpeyaṣya. tvachā tiktā kasāyā madhurā tathā." (Sva.)

Bark – Anthelmintic, Astringent, Bitter, Depurative, Haemostatic, Sweet, Vulnerary.

Medicinal Uses: Bark - Colic, Diarrhoea,Gout, Haemorrhages, Helminthiasis, Leprosy, Stomatitis, Syphilis, Ulcers, Wounds.

Ethno- Medicinal Uses: Leaf, Root, Seed - Diabetes (Ghosh, 2006).

TABERNAEMONTANA HEYNEANA Wall. (APOCYNACEAE)

Syn.: *T. alternifolia* L.; *Ervatamia heyneana* (Wall.) Cooke.; *Ervatamia alternifolia* (L.) Almedia.

Regional & Other Names : *English* – East Indian rose bay, Wax flower; *Sanskrit* – Sveta-kutaja.

Distribution: Common in open forests areas in Konkan & Western ghats.

Taxonomic Description: A shrub or small deciduous tree.

Medicinal Parts: Wood, Milky juice, Root & Stem bark.

Chemical Constituents: Root - Coronaridine, Voacangine, Ibogamine.

Fruits - Lupeol-O Ac and β -Amyrin-OAC.

Leaves - Alkaloid-Isocoacristine.

Medicinal Properties:

श्वेस्तु कुट्जस्तिक्तः कटुश्वोष्णोऽग्निदीपकः ।
पाचकस्तुवरो रूक्षो ग्राहको रक्तदोषहा ।। (नि. र.)

Ṣvetaṣtu kuṭjastikaḥ katuṣcoṣṇa agnidīpakaḥ
Pācakaṣluvarō rūksograhako raktadosahā. (Ni. Ra.)

Root- Acrid, Alexipharmic, Astringent, Bitter, Digestive.

Medicinal Uses: Root- Biliousness, Diseases of blood.

TARAXACUM OFFICINALE Weber. (ASTERACEAE)

Regional & Other Names : *Bengali* – Pitachumki; *English* – Common dandelion; *Gujarati* - Pathardi; *Hindi* – Dudhal; *Kannada* – Kadusevanti; *Malayalam* – Dugddhapheni; *Marathi* –Undarkani; *Sanskrit* – Dudhapheni; *Tamil* – ; *Telugu* – Patri.

Distribution: Throughout India.

Taxonomic Description: Perennial hebs, with Milky juice & Yellow flowers.

Medicinal Parts: Whole plant.

Chemical Constituents: Root contains - Inulin, pectin, Sugar, Levulin. The milky juice contains a bitter amorphous Principle- Taraxacin, A Crystalline Principle-Taraxacerin; Also Potassium & Calcium salts, Resinoid and Glutinous bodies.

Medicinal Properties:

दुग्धफेनी कटुस्तिक्ताऽशिशिरा विषनाशिनी ।
व्रणापसारणी रूच्या युक्त्या चैव रसायनी ।। (रा. नि.)

Dughapheni katuṣtiktāaśiśira visanāśinī
Vṛanāpasāranī rūcyā yuktyā caīva rasāyanī. (Rā. Ni.)

Plant- Acrid, Anthelmintic, Anti-Inflammatory, Aperient, Bitter, Depurative, Digestive, Diuretic, Febrifuge, Liver tonic, Sudorific, Stimulant, Stomachic, Thermogenic, Vulnearay, Tonic.

Medicinal Uses: Plant - Colic, Constipation, Dyspepsia, Fever, Flatulence, General debility, Gout, Inflammation, Insomnia, Jaundice, Leprosy, Skin diseases, Strangury, Tuberculosis, Verminosis.

Ethno- Medicinal Uses: Whole plant – Blisters, Wounds (Jain,1984); Dysentery (Singh *et al.*, 1980).

Root - Blood purifier, Gastric ulcers (Uniyal & Chauhan, 1973); Kidney diseases (Jain,1984; Singh *et al.*, 1980; Dhasmana, 1986; Gaur *et al.*, 1983; Srivastava *et al.*, 1981); Liver complaint (Singh *et al.*, 1980; Dhasmana, 1986; Srivastava *et al.*, 1981); Tonic (Gaur *et al.*, 1983); Headache, Vertigo (Srivastava *et al.*, 1981); Dyspepsia, Jaundice, Dropsy, Skin diseases (Kapur & Nanda, 1992); Joints pain (Ara & Naqshi, 1992); Swelling (Ara & Naqshi, 1992); Liver tonic (Chhetri, 2005); Tonic, Diuretic (Chaurasia *et al.*, 2003); Mouth blisters (Kala & Manjrekar, 2003).

Stem, Flowers - Mouth blisters (Kala & Manjrekar, 2003).

TERMINALIA PANICULATA Roth. (COMBRITACEAE)

Regional & Other Names : *English* – Flowering murdah; *Kannada* – Hanalu; *Malayalam* – Pumarutu; *Marathi* – Kinjal; *Sanskrit* – Asvakarnah; *Tamil* – Pumarutu; *Telugu* – Nimiri.

Distribution: In the Western & Eastern Ghats, upto 1,200m.

Taxonomic Description: Large deciduous trees, with Reddish Brown flowers & Reddish Brown-winged fruits.

Medicinal Parts: Bark.

Chemical Constituents: Bark & Fruit contain Tannin.

Medicinal Properties:

कषायतिक्तः कफहा व्रणसंरोपणः परः ।
पाण्डुमेहप्रशमनः कुष्ठहा वाजिकर्णिका ।। (म. नि.)

Kasāyatiktaḥ kaphahā vṛaṇasaṃrōpaṇaḥ paraḥ
Pāṇdumehapṛaśamanaḥ kuṣthahā vājikaṛṇikā. (Ma. Ni.)

Bark - Astringent, Bitter, Cardiotonic, Cooling, Depurative, Diuretic, Expectorant, Tonic, Vulnerary.

Medicinal Uses: Bark - Bronchitis, Cardiac debility, Cough, Diabetes, Leprosy, Skin diseases, Strangury, Ulcers, Wounds.

Ethno- Medicinal Uses: Bark - Jaundice, Mumps (Sasidharan & Augustine, 2006).

THALICTRUM FOLIOLOSUM DC. (RANUNCULACEAE)

Regional & Other Names : *Bengali* – Gurbiani; *Hindi* – Pilazari; *Sanskrit* – Mamira.

Distribution: Throughout the Himalaya, 5,000-8,000 ft., Khasia hills, 4,000-6,000 ft., Burma.

Taxonomic Description: Perennial herbs, with Green or Violet flowers & Achene fruits.

Medicinal Parts: Roots.

Medicinal Properties:

तिक्तमूला कटुस्तिक्ता रूक्षोष्णा दीपनी सरा ।
चक्षुष्या पाचनी रूच्या कफपित्तहरा मता ।। (द्र. गु. वि.)

Tiktamūlā katustiktā ruksōsnā dīpanī sarā
Caksusyā pācanī rucyā kaphapiṭtaharā matā. (Dra. Gu. Vi.)

Root - Diuretic, Purgative, Bitter.

Medicinal Uses: Conjunctivitis, Leucoderma.

Root- Ophthalmia, Dyspepsia.

Ethno- Medicinal Uses: Root - Abdominal pain, Blood purifier, Boils (Uniyal & Chauhan, 1973); Eczema (Shah & Joshi, 1981); Eye diseases (Uniyal & Chauhan, 1973; Shah & Joshi, 1981; Sharma *et al.*, 1979; Sinha *et al.*, 2006); Fever, Leucoderma, Piles (Sharma *et al.*, 1979); Piles, Earache, Toothache (Malhotra & Balodi, 1984); skin diseases, sores (Sinha *et al.*, 2006).

Leaf- Eye diseases, skin diseases, sores (Sinha *et al.*, 2006).

TINOSPORA MALABARICA (Lam.) Miers. (MENISPERMINACEAE)

Syn.: *T. sinensis* (Roxb.) Merr.

Regional & Other Names : *Bengali* –Padmagaluncha; *Hindi* –Giloe; *Kannada* – Sudarsnaballi; *Marathi* – Gulvel; *Sanskrit* – Kandodbhava-guruci; *Tamil* – Potchindil.

Distribution: Konkan, Karnataka, Tamil Nadu, Bengal, Orissa.

Taxonomic Description: A gaint deciduous climber.

Medicinal Parts: Leaves, Stem.

Chemical Constituents: Root & Stem contains Starch, a Bitter principle and a tace of berberine.

Medicinal Properties:

कन्दोद्भवा गुरूचि च कटुष्णा सन्निपातहा ।
विषघ्नी ज्वरभूतघ्नि वलीपलितनाशिनी ।। (रा. नि.)

Kandodbhavā guruci ca katusnā sannipātahā
Visaghnī jvarabhutaghni balīpalitanāsinī (Ra. Ni.)

The plant has tonic properties.

Medicinal Uses: Tonic.

Ethno- Medicinal Uses: Leaf - Eye diseases (Singh *et al.*, 1987).

Stem - Lactation (Srivastava & Varma, 1981); Urinary complaint (Kumar *et al.*, 1987).

Root / Stem bark - Gonorrhoea. (Purkayastha & Nath, 2006).

TODDALIA ASIATICA (LIN.) LAM. (RUTACEAE)

Syn.: *T. aculeata* Pers.

Regional & Other Names : *Bengali* – Dahan; *English* – Wild orange tree; *Hindi* – Dahan; *Kannada* – Kadumaraju; *Malayalam* – Kara; *Marathi* – Manger; *Sanskrit* – Dahana; *Tamil* – Milagaranai; *Telugu* – Mirapokauda.

Distribution: Konkan, Deccan, S.M. county, N. Kanara, Madras Presidency, Ceylon, Kumaon between 1,000 and 4,500 ft.

Taxonomic Description: An evergreen climber, with pale Greenish, Yellow flowers & globose, Yellowish fruits.

Medicinal Parts: Roots, Leaves, Flowers, Fruits.

Chemical Constituents: Root - Coumarins as Toddanol, Toddanone, toddasin, Pimplinellin, Iso-Pimpinellin; Alkaloids as Norchelerythrine, Dihydronitidnine, Skimmianine, Toddaline, Toddalinine, Chelerythrine, Nitidine, β -amyrine.

Leaves - Toddaline, Toddalinine, Toddalolactone, Pimpinellin, Isopimpinellin, Resins.

Medicinal Properties:

रक्तस्तु काञ्चनः शीतः सरोह्यग्नि प्रदीपनः ।
तुवरो ग्राहकः प्रोक्तः कफपित्तव्रणकृमीन् ।। (नि. र.)

Raktaṣtu kāṇcanaḥ śītaḥ saro hyagni pradīpanaḥ
Tuvarō gṛāhakaḥ pṛoktaḥ kaphapiṭtavraṇakṛmīṇ. (Ni. Ra.)

Root - Acrid, Antiperiodic, Antipyretic, Aromatic, Astringent, Bitter, Carminative, Constipating, Diaphoretic, Digestive, Expectorant, Sweet, Vulnerary.

Medicinal Uses: Roots - Bronchitis, Colic, Cough, Diarrhoea, Dyspepsia, Epilepsy, Flatulence, General debility, Gonorrhoea, Intermittent Fever, Malaria, Nausea, Paralysis, Wound.

Ethno- Medicinal Uses: Leaf - Bodyache(Abraham, 1981); Burns (Panda & Das, 2003); Pains in the bowels (Rosakutty *et al.,* 2003).

Root bark – Stomachache (Sharma *et al.,* 1979); Toothache (Banerjee, 1977).

Stem - Vomiting (Kshirsagar & Singh, 2003).

Whole plant - Burning sensation inside stomach (Binu *et al.,* 2003).

Root - Malaria (Purkayastha & Nath, 2006); Cuts, Wounds (Udayan *et al.,* 2006).

TRACHYSPERMUM AMMI (L.) SPRAGAE. (APIACEAE)

Syn.: *Carum copticum* Benth & Hook.

Regional & Other Names : *Bengali* – Javan; *English* – Carum; *Gujarati*-Ajma; *Hindi* – Ajwain; *Kannada* – Oma; *Malayalam* – Ayanudakan; *Marathi* – Onva; *Sanskrit* – Ajawodika; *Tamil* – Omam; *Telugu* – Vamu.

Distribution: Throughout India.

Taxonomic Description: Annual herbs, with whitish or pinkish flowers and muricated fruits.

Medicinal Parts: Fruits, Leaves, Oil.

Chemical Constituents: Fruits contain - An aromatic Volatile oil, Stearoptene, Cumene, Terpene as Thymine; Phenolic Glucosides; Thymol, p-cymene, Stearoptin, Cymon, Terpene.

Medicinal Properties:

यवानी कटुतीक्ष्णोष्णा वातश्लेष्मद्विजायमान् ।
हन्ति गुल्मोदरं शूलं दीपयत्याशु चानलम् ।। (ध. नि.)

Yavāni katutiksnosnā vātaślesmadvijāyamāṇ
Hatṇi gulmodaraṃ shulaṃ dipayatyāshu chānlaṃ (Dha. Ni.)

Fruits - Anthelmintic, Antidiarrhoeal, Appetiser, Digestive, Stomachic, Tonic.

Leaves - Anthelmintic, Appetiser, Digestive, Heart tonic, Stimulant, Stomachic, Tonic.

Medicinal Uses: Fruits - Abdominal Colic & Pain, Cholera, Diarrhoea, Dyspepsia, Flatulence, Heart tonic, Helminthiasis, Indigestion, Rheumatism, Uritcaria.

Leaves - Abdominal pain, Cholera, Colic, Diarrhoea, Flatulence, Heart diseases, Helminthiasis, Indigestion, Rheumatism, Urticaria.

Ethno- Medicinal Uses: Seed - Boils, Cuts, Colic (Saxena & Vyas, 1986; Diarrhoea (Shukla *et al.,* 1992); Indigestion, Stomachache, Dyspepsia (Jadeja *et al.,* 2006); Diarrhoea & Dysentery (Kataria & Gauchan, 2006).

Leaves - Night blindness (Jadhav, 2006).

Flowers & Fruits - Cough, Cold, Headache (Deokule, 2006).

Fruit- Abortifacient, pneumonia (Singh & Pandey, 1980).

TRACHYSPERMUM ROXBURGHIANUM (DC.) Craib (APIACEAE)

Syn.: *T. involucratum* Woltt. non Marie; *Carum roxburghianum* Benth. ex Kurz.

Regional & Other Names : *Bengali* –Ajmud; *English* – Ajowan; *Gujarati* - Ajmod; *Hindi* – Ajmud; *Kannada* – Ajmodhavoma; *Malayalam* – Ayamodakam; *Marathi* – Ajmoda-vova; *Sanskrit* – Ajoda; *Tamil* – Asamta-vomam; *Telugu* – Ajumoda.

Distribution: Cultivated throughout India.

Taxonomic Description: Annual herbs, with white flowers and cremocarp fruits.

Medicinal Parts: Fruits.

Chemical Constituents: *Fruits contain* □-terpinene, d- linalool, di-terpienol, thymohydroquinone, thymol, di-pipretione, cucminic acid, cumminaldehyde, d-limonene and dipentene mixture.

Medicinal Properties:

अजमोदा च शूलध्नी तिक्तोष्णा कफवातजित् ।
हिक्काध्मानारुचिर्हन्ति कृमिजिद्वह्निदीपनी ।।(ध. नि.)

Ajamoda ca sulaghni tiktosna kaphavatajit
Hikkadhmanarucirhanti krimijidvahnidipani (Dha. Ni.)

Fruits – Acrid, Anthelmintic, Bitter, Carminative, Digestive, Stomachic, Tonic.

Medicinal Uses: Fruits – Asthma, Diarrhoea, Dyspepsia, Flatulence, Hiccough, vomiting.

TRAGIA INVOLUCRATA L. (EUPHORBIACEAE)

Regional & Other Names : *Bengali* – Bichati; *English* – Indian stinging Nettle; *Gujarati* – Motikhajavanini Bel; *Hindi* – Barahanta; *Kannada* – Kiriturachi; *Malayalam* – Kottituva; *Marathi* – Khajkolati; *Sanskrit* – Dhumpushpa; *Tamil* – Kancori; *Telugu* – Dulagudi.

Distribution: Throughout India.

Taxonomic Description: Perennial evergreen unisexual climber, with capsules fruits & globose, Smooth seeds.

Medicinal Parts: Roots, Leaves.

Chemical Constituents: Linoleic acid is a main constituent of the fatty oil.

Medicinal Properties:

दुरालंभा स्वादुशीता तिक्ता वातविनाशिनी ।
विषमज्वरतृड्च्छर्दिमोहविनाशिनी ।। (ध. नि.)

Durālaṃbhā ṣvāduśītā tiktā vātavināśinī
Visamjvaratṛdcchaṛdimehamohavināśinī. (Dha. Ni.)

Root - Acrid, Alterant. Antiperiodic, Bitter, Cooling, Depurative, Diaphoretic, Diuretic, Sweet.

Medicinal Uses: Root - Blood impurities, Diabetes, Dipsia, Gastropathy, Giddiness, Guinea worms, Haemorrhoids, Venereal diseases, Vomiting.

Ethno- Medicinal Uses: Fruits - Baldness (Mishra & Sahu, 1984); Skin diseases (Saxena & Dutta, 1975).

Root - Lactation (Chaudhuri *et al.,* 1976); Pain (Mishra & Sahu, 1984); Snake bite (Saxena & Dutta, 1975).

Leaf - Dog bite (Majumdar *et al.,* 1978); Headache (Singh,2003).

Stem - Scabies (Majumdar *et al.,* 1978).

Seed - Urinary infection (Arinathan *et al.,* 2003).

Whole plant - Fever (Prasad *et al.,* 2003).

TRIANTHEMA PORTULACASTRUM L. (AZOIACEAE)

Regional & Other Names : *Bengali* – Sabuni; *English* – Horse purslane; *Hindi* – Sabuni; *Marathi* – Pundharighentuli; *Sanskrit* – Chiratika; *Tamil* – Shavalai; *Telugu* – Ambatimadu.

Distribution: Throughout India.

Taxonomic Description: Succulent, prostrate herbs, with spathulate leaves. Flowers white or pink, solitary; stamens 10. capsules mitrifrom.

Medicinal Parts: Roots, Leaves, Fruits, Seeds.

Chemical Constituents: Protein, fat, carbohydrate, crude fibre, calcium, phosphorus, Iron, ascorbic acid, carotene, oxalates.

Medicinal Properties:

कठिल्लकं हिमं तिक्तं विपाके कटुकं लघु ।
संग्राहि वातलं पित्तकफशोणितनाशनम् ।। (कै. नि.)

Kathillakaṃ himaṃ tiktaṃ vipāke katukaṃ laghu
Saṇgrāhi vātalaṃ pittakaphaśonitnāsanaṃ (Kai. Ni.)

Plant - The plant is Alexiteric, Analgesic, Alterative, Bitter, Hot, Laxative, Stomachic.

Medicinal Uses: Plant - Anaemia, Ascites, Bronchitis, Heart diseases of the blood, Inflammation, Piles,

Root - Night blindness.

Ethno- Medicinal Uses: Root - Jaundice (Sekhawat & Sunil Anannd, 1984); Boils (Shukla *et al.,* 1992); Constipation (Jadeja *et al.,* 2006).

Leaf- Liver complaint (Sudhakar & Rolla, 1985; Sinha *et al.,* 2006); Jaundice (Shukla *et al.,* 1992); Seasonal fever (Sharma *et al.* 2006).

Whole plant - Kidney disorders (Thomas & Britto, 2003); Abdominal disorders, diarrhoea (Basak, 2006); Eyes infection (Sharma *et al.,* 2006).

TRICHODESMA INDICUM (L.) R. Br. (BORAGINACEAE)

Regional & Other Names : *Bengali* – Chotokulpa; *Gujarati* - Undhaphuli; *Hindi* – Chhota kulpa; *Kannada* – Adhomukhi; *Marathi* – Chhotaphulva; *Sanskrit* – Surasa; *Tamil* – Kazuthai-tumbai; *Telugu* – Gusva-gutti.

Distribution: Throughout India.

Taxonomic Description: An erect or diffuse, much branched, Grey-Villous hispid herb. Leaves mostly sessile lanceolate or cordate-lanceolate. Flowers pale-Blue or Violet or White with Brown throat. Nutlets rough on the inner face, obscurely margined.

Medicinal Parts: Whole plant.

Chemical Constituents: *Leaves contain* - Hexacosane, ethyl hexacosanoate, ethyl 21, 24-hexacosadienoate.

Medicinal Properties-

अधःपुष्पी च चक्षुष्या मूढगर्भापकर्षणी ।
मोदयन्ती कटूष्णा च विषूची कफवातजित् ।। (सो. नि.)

Adhaḥpuṣpī ca caksuṣyā mudhagarbhapakaṛvsaṇī
Modyaṇtī katuṣṇā ca visucī kaphavātajiṭ. (So. Ni.)

Plant - Acrid, Alexeteric, Anodyne, Anti-Inflammatory, Bitter, Carminative, Constipating, Depurative, Diretic, Emollient, Febirufuge, Ophthalmic, Thermogenic,

Medicinal Uses: Plant - Diarrhoea, Dysentery, Dysmenorrhoea, Dyspepsia, Fever, Inflammation, Leprosy, Skin diseases, Strangury,

Ethno- Medicinal Uses: Fruits - Swelling (Sharma *et al.,* 1979).

Whole plant - Cough, Fever (Kapur & Nanda, 1992); Boils (Srivastava & Varma, 1981).

Leaf - Stomach upset (Ganesan *et al.,* 2006); Eczema (Saxena & Vyas, 1983; Sinha *et al.,* 2006); Boils, Stomach pain (Sinha *et al.,* 2006).

Root - Dysentery in children, Swellings in joints (Thomas & Britto, 2003); Stomachache (Jain 1963; Sinha *et al.,* 2006); Eczema, Boils (Sinha *et al.,* 2006).

TRICHOSANTHES CUCUMERINA L. (CUCURBITACEAE)

Syn.: *T. anguina* L.

Regional & Other Names : *Bengali* – Banpatol; *Gujarati* - Kadwamparwal; *Hindi* – Jungli chachinda; *Kannada* – Kiripodla; *Malayalam* – Patavalam; *Marathi* – Parula; *Sanskrit* – Amritahala; *Tamil* – Kabtupeypudal; *Telugu* – Patolamu.

Distribution: Cultivated thoughout India.

Taxonomic Description: A climbing or spreading, scabrous, stout herb. Tendrils divided. Leaves usually 5-lobed or angled; Margins denticulate. Flowers White, with fimbriase petals. Male peduncles paired. Female flowers solitary. Fruits Yellow, pointed.

Medicinal Parts: Whole plant.

Chemical Constituents: Leaves - β-Sitosterol Rits gloucoside and luteolin-7-glucoside.

Seed oil - Punicic acid, Oleic acid, Linoleic acid, Palmitic acid, stearic acid, arachidic acid, Cucurbitacin-B.

Medicinal Properties:

पटोलपत्रं पित्तघ्नं वल्ली चास्य कफापहा ।
फलं त्रिदोषशमनं मूलं चास्य विरेचनम् ।। (ध. नि.)

Patolapaṭraṃ piṭtagnaṃ vaḷlī cāṣya kaphāpahā
Phalaṃ ṭridosaśamanaṃ mūlaṃ cāṣya virecanaṃ. (Dha. Ni.)

Roots / Seeds - Anthelmintic, Cooling, Febrifuge, Purgative, Vermifuge, Tonic.

Leaves - Cooling.

Fruits - Anthelmintic, Carminative, Cooling, Digestive, Febrifuge, Purgative, Sweet, Tonic.

Medicinal Uses: Roots / Seeds - Syphilis, Verminosis.

Fruits - Anorexia, Burning sensation, Constipation, Dipsia, Dyspepsia, Fever, Flatulence, General weakness. Helminthiasis.

Ethno- Medicinal Uses: Fruits - Jaundice (Maheshwari *et al.,* 1986).

Root - Stomachache (Krishna & Singh, 1987).

Seed - Stomachache, Worm (Jadeja *et al.,* 2006).

Leaves, Stem Bark - Dandruff (Ghosh, 2006).

TRICHOSANTHES LOBATA Roxb. (CUCURBITACEAE)

Regional & Other Names : *English* – Bitter snake gourd; *Hindi* – Jangli cicinda; *Kannada* – Kiripodla; *Malayalam* – Peppatolam; *Sanskrit* – Patola; *Tamil* – Peppudal; *Telugu* – Kommupotla.

Distribution: Throughout India.

Taxonomic Description: Annual monoecious tendriller climbers, with White flowers & Berries fruits.

Medicinal Parts: Whole plant.

Medicinal Properties-

पटोलं कटुकं तिक्तमुष्णं पित्तविरोधि च ।
कफासृक्कण्डुकुष्ठानि ज्वरदाहौ च नाशयेत् ।। (ध. नि.)

Patolaṃ katukaṃ tiktamuṣṇaṃ piṭtavirodhi ca
Kaphāsṛkkaṇdukuṣthāni jvaradāhaū ca nāśayet. (Dha. Ni.)

Plant - Acrid, Alexeteric, Anodyne, Anthelmintic, Anti-Inflammatory, Appetiser, Bitter, Carminative, Depurative, Digestive, Emollient, Expectorant, Laxative, Stomachic, Thermogenic, Tonic, Vulnerary.

Medicinal Uses: Plant - Anorexia, Asthma, Boils, Bronchitis, Colic, Cough, Dipsia, Dyspepsia, Flatulence, General debility, Helminthiasis, Hyperacidity, Inflammation, Intermittent Fever, Jaundice, Leucoderma, Pruritus, Ulcer, Wounds.

TRICHOSANTHES TRICUSPIDATA Lour. (CUCURBITACEAE)

Syn.: *T. bracteata* (Lam.) Voigt.; *T. palmata* Roxb.

Regional & Other Names : *Bengali* –Makal; *Gujarati* -Ratanindrayan; *Hindi* – Lal indrayan; *Kannada* – Avagudehannu; *Malayalam* – Kakkattonti; *Sanskrit* – Mahakasa; *Tamil* – Korattai; *Telugu* – Avaduta.

Distribution: Throughout India.

Taxonomic Description: Woody tendril climbers, with unisexual flowers & Orange streaked Red globose fruits.

Medicinal Parts : Roots, Fruits.

Chemical Constituents: Root - Trichotetrol.

Leaf - Cyclotrichosanthol, Cycloeucalenol.

Medicinal Properties:

काकनासा कषायोष्णा कटुका रसपाकयोः ।
कफघ्नी वामनी तिक्ता शोफार्शः श्वित्रकुष्ठहृत् ।। (भा. प्र.)

Kākanāsā kasāyoṣṇā katukā rasapākayoḥ
Kaphaghnī vāmanī tiktā śophāṛśaḥṣviṭrakuṣthahṛṭ. (Bhā. Pṛa.)

Root - Bitter, Astringent, Acrid, Thermogenic, Carminative, Emetic, Abortifacient, Purgative.

Medicinal Uses: Roots - Hemicrania, Carbuncles, Gonorrhoea.

Fruits - Rhinitis, Inflammation, Asthma, Epilepsy, Leprosy.

Ethno- Medicinal Uses: Root - Diarrhoea (Joshi *et al.,* 1980); Fever (Shah & Joshi, 1981); Pneumonia (Saxena *et al.,* 1981); Rheumatism (Tribedi *et al.,* 1985); Snake bite (Audichya *et al.,* 1983).

Seed - Dysentery (Sen *et al.,* 1985); Epilepsy (Ravikumar & Sankar, 2003).

Fruits - Asthma, Earache (Ganesan *et al.,* 2006); Asthma (Basak, 2006).

TRIDAX PROCUMBENS L. (ASTERACEAE)

Regional & Other Names : *Bengali* – Tridaksha; *English* – Coat buttons; *Kannada* – Gabbu sanna; *Tamil* – Vel; *Telugu* – Balapaku.

Distribution: Throughout India.

Taxonomic Description: Straggling herbs, with ovate-lanceolate leaves & yellow heads terminating long peduncles. Achenes silky.

Medicinal Parts: Whole plant.

Chemical Constituents: Leaf - Fumaric acid.

Flowers - Glucoluteolin, Luteolin, Isoquercetin, Quercetin.

Medicinal Properties: Plant - Hypotensive.

Leaf juice- Insecticidal.

Medicinal Uses:

Ethno- Medicinal Uses: Leaf - Antiseptic (Malhotra & Moorthy, 1973; Sharma *et al.,* 1979; Sinha *et al.,* 2006); boils (Sebastian & Bhandari, 1984; Dwivedi & Pandey, 1992; Pawar & Patil, 2006; Sinha *et al.,* 2006); cuts & wounds (Nagendra Prasad & Abraham, 1984; Sudhakar & Rolla, 1985; Ramachandran & Nair, 1981; Maheshwari *et al.,* 1986; Upadhye *et al.,* 1986; Goel & Mudgal, 1988; Deokule & Magdum, 1992; Rosakutty *et al.,* 2003; Ganesan *et al.,* 2006; Kiurba *et al.,* 2006); diarrhoea & dysentery (Sebastian & Bhandari, 1984; Mukhopadhyay, 1992; Subramaniam, 2003); eye diseases (Goel *et al.,* 1984); leprosy (Goel & Mudgal, 1988); scorpion bite (Jain & De, 1966); toothache (Sudhakar & Rolla, 1985; Dhasmana, 1986); Bronchial Troubles (Mukhopadhyay, 1992); Blood clotting (Sur *et al.,* 1992); Cough in children (Deokule & Magdum, 1992); Eczema (Bhatt & Sabnis, 1987; Pawar & Patil, 2006); Cuts (Chaudhury & Neogi, 2003; Kumar & Chauhan,2005; Deokule, 2006; Pawar & Patil, 2006); Piles (Dwivedi, 2003; Dwivedi & Pandey, 1992); Wounds (Rajendran *et al.,* 2003; Deokule, 2006; Udayan, *et al.,* 2006); Burns (Deokule, 2006); Headache, Eye inflammation (Sharma et al. 2006).

Whole plant - Urinary stone bladder (Pawar & Patil, 2006); Wounds & Boils (Panda & Das, 2003); Bleeding (Sharma *et al.* 2006); Spermatorrhoea (Prakash & Singh, 2006); Ulcers (Pandey *et al.,* 1981).

Seeds - Diarrhoea & Dysentery (Kataria & Gauchan, 2006).

Root- Skin diseases (Jain & De, 1966).

TRITICUM AESTIVUM L. (POACEAE)

Regional & Other Names : *Bengali* – Gom; *English* – Bread wheat; *Gujarati* - Gawn; *Hindi* – Gehun; *Kannada* – Godhi; *Malayalam* – Gendum; *Marathi*–Gahum; *Sanskrit*-Arupa; *Tamil* – Godumai; *Telugu* – Goodhumalu.

Distribution: Largely cultivated throughout India.

Taxonomic Description: Medium-sized annuals. Inflorescence a terminal distichous spike; Spikelets solitary, awned. Grains elliptic-oblong.

Medicinal Parts: Grains.

Chemical Constituents: Grians - Protein, Crude fibres, Sugars, Fat, Minerals, Vitamins, Enzymes.

Medicinal Properties:

मधुना जीर्णगोधूमं कफशूले प्रयोजयेत् ।। (भा. प्र.)

Madhunā jīrṇagodhūmaṃ kaphasūle prayojayeṭ. (Bhā. Pṛa.)

Grains – Aphrodisiac, Appetiser, Demulcent, Emollient, Laxative,

Refrigerant, Rejuvenating, Sweet, Vulnerary.

Medicinal Uses: Grains - Anorexia, Boils, Constipation, Flatulence, General debility, Mernorrhagia, Pruritus, Skin diseases, Ulcers.

Ethno- Medicinal Uses: Flour - Swelling, Sprain, Boils, (Deokule, 2006).

Grains - Eczema (Kumar & Yadav, 2006); Gout (Jadeja *et al.,* 2006).

Leaf - To make strong & silky hairs (Ghosh, 2006).

TRIUMFETTA RHOMBOIDEA Jacq. (TILIACEAE)

Syn.: *T. bartramia* L.; *T. angulata* Lam.

Regional & Other Names : *Bengali* – Ban-okra; *English* – Spiny cockle burr; *Gujarati* - Jhipato; *Hindi* – Chikti; *Kannada* – Kadubende; *Marathi* – Jhinjudi; *Sanskrit* – Jhinjharita; *Tamil* – Ottupullu; *Telugu* – Tutturubenda.

Distribution: Throughout India.

Taxonomic Description: Stout herbs, with ovate-rounded, 3-lobed & entire Leaves. Flowers yellow, with 5-10 stamens. Capsules pubescent, spines glabrous.

Medicinal Parts: Seeds, Bark.

Chemical Constituents: Plant contains - 4-Hydroxyisoxazole and Triumboidin.

Medicinal Properties:

झिन्झिरीटा कटुः शीता कषाया चातिसारजित् ।
वृष्या सन्तर्पणी वष्या महिषीक्षीरवर्द्धनी ।। (रा. नि.)

Jhiṇjhirītā katuḥ śītā kasāyāh cātisārajiṭ
Vṛṣyā saṇtaṛpaṇī baḷyā mahisīksīravaṛddhinī. (Rā. Ni.)

Root - Acrid, Aphrodisiac, Bitter, Cooling, Galactagogue, Styptic, Tonic.

Medicinal Uses: Root - Dysentery.

Ethno- Medicinal Uses: Whole plant - Antiferility (Shah *et al.*, 1983)

Leaf - Diarrhoea (Sebastian & Bhandari, 1984); stomachache (Kumar *et al.*, 1987; Sinha *et al.*, 2006); impotency (Shah *et al.*, 1983); Diarrhoea, Dysentery (Sinha *et al.*, 2006).

Root - Boils, Inflammed eyelids (Upadhyay & Singh, 2005); Dysentery (Subramaniam, 2003; Sinha *et al.*, 2006); Diarrhoea, Stomachache (Sinha *et al.*, 2006).

Seeds - Diarrhoea & Dysentery (Kataria & Gauchan, 2006).

TYPHA ELEPHANTINA Roxb. (TYPHACEAE)

Syn.: *T. angustifolia* Watt. non L.

Regional & Other Names : *Bengali* – Hogla; *English* – Elephant grass; *Gujarati* - Ghabajarin; *Hindi* – Mothitrina; *Kannada* – Apu; *Marathi* – Rambdna; *Sanskrit* –Gundra; *Tamil* – Chamba; *Telugu* – Kandra.

Distribution: Throughout India.

Taxonomic Description: Erect, rhizomatous herbs, with spongy linear leaves flat above and convex beneath. Spikes cylindrical, male flowers above, female below and up to 3 cm long gap in between.

Medicinal Parts: Whole plant.

Chemical Constituents: Seed yield a fatty oil.

Medicinal Properties:

एरका शिशिरा वृष्या चक्षुष्या वातकोपनी ।
मूत्रकृच्छ्राशमरी दाहपित्तशोणितनाशिनी ।। (कै. नि.)

Erkā śiśirā vṛiṣyā caksuṣyā vātkopanī
Mūṭrakṛucchāśhmarī dāhapiṭtśoṇitnāśinī. (Kaī. Ni.)

Whole plant - Cooling, Ahprodisiac.

Medicinal Uses: Whole plant - Burning sensation, Gonorrhoea, Inflammation, Leprosy, Splenic enlargement, Strangury.

Ethno- Medicinal Uses: Fruit- Stop bleeding (Shah, 1984).

Root stalk & Inflorescnce - Wounds (Deokule & Magdum, 1992).

Roots - Permanent sterility in female (Bhogaonkar & Kadam, 2006).

Seed - Facial cosmetic (Rajendra & Aswal, 2003).

Flowers - Wounds (Deokule, 2006).

UVARIA NARUM (Dunal.) Wall. ex Wight. (ANNONACEAE)

Regional & Other Names : *Kannada* – Kariballi; *Malayalam* – Kuril; *Sanskrit* – Nilavalli; *Tamil* – Puliccan.

Distribution: In the forests of Western ghats upto 1,200 m.

Taxonomic Description: Woody stellately pubescent straggling shrub with dark Bluish Green leaves & Reddish flowers.

Medicinal Parts: Roots, Leaves.

Chemical Constituents: Essential oil.

Medicinal Properties:

नीलवल्याः मूलं कषायंमधुरं स्निग्धं सुशीतलम् ।। (स्व.)

"Nīlavaḷlyāḥ mūlaṃ kasāyaṃ madhuraṃ ṣnigdhaṃ suśītalaṃ."
(Ṣva.)

Roots - Aromatic, Astringent, Depurative, Emollient, Febrifuge. Fragrant, Stomachic, Sweet.

Leaf - Anti-Inflammatory, Astringent, Cooling, Febrifuge, Sour, Sweet.

Medicinal Uses: Roots - Eczema, Erysipelas, Gastropathy, Jaundice, Pruritus.

Leaf – Fever, Inflammation, Jaundice.

Ethno-Medicinal Plants: *Leaf* - Urinogenital diseases (Bhandari & Chandrashekar, 2002).

VALERIANA JATAMANSI Jones. (VALERIANACEAE)

Syn.: *V. wallichii* DC.

Regional & Other Names : *Bengali* – Mushkbala; *English* – Indian valerian; *Hindi* – Tagar; *Kannada* – Tagara; *Malayalam* – Takaram; *Sanskrit* – Tagarah; *Tamil* – Takaram; *Telugu* – Tagara.

Distribution: In the Himalayas upto 3,000 m.

Taxonomic Description: Hairy perennial herbs, with White unisexual flowers & Compressed oblong fruits.

Medicinal Parts: Roots.

Chemical Constituents: Roots-Valerianic acid, Valepotriates, Valerosidatum glucoside; Essential oil contains Sesquiterpenes, Valerianic acid, Terpene alcohol, Bornyl Esters of Formic, Acetic, Butyric & Isovaleric acids.

Medicinal Properties:

तगरं शीतलं तिक्तं दृष्टिदोषविनाशनम् ।
विषार्त्तिशमनं पथ्यं भूतोन्मादभयापहम् ।। (रा. नि.)

Tagaraṃ śītalaṃ tiktaṃ dṛṣtidosavināśanaṃ
Visāṛttiśamanaṃ pathyaṃ bhūtoṇmādabhayāpahaṃ. (Rā. Ni.)

Roots - Acrid, Alexiteric, Anodyne, Aphrodisiac, Astringent, Bitter, Carminative, Digestive, Diuretic, Emollient, Expectorant, Febrifuge, Laxative, Stomachache, Thermogenic, Tonic, Vulnerary.

Medicinal Uses: Roots - Amenorrhoea, Ascites, Asthma, Bronchitis, Colic, Constipation, Dysmenorrhoea, Dyspepsia, Epilepsy, Flatulence, General debility, Intermitten Fever, Jaundice, Leprosy, Seminal weakness, Skin diseases, Strangury, Ulcers, Wounds.

Ethno- Medicinal Uses: Rhizome - Epilepsy (Shah, 1982; Gupta, 1981); Hysteria (Gupta, 1981); Urinary complaint (Shah & Joshi, 1981); Hair oil (Uniyal, 1968).

Root - Hysteria (Arya & Agarwal, 2006).

VATERIA INDICA L. (DIPTEROCARPACEAE)

Regional & Other Names : *Bengali*–Kundaro; *English* – Piney varnish; *Gujarati* – Safed damar; *Hindi* – Chandras; *Kannada* – Rala; *Malayalam* – Payani; *Marathi* – Raal; *Sanskrit* – Ajakaran; *Tamil* – Painimaram; *Telugu* - Telladamara.

Distribution: In the Western ghats, upto 1,200 m. along the streams.

Taxonomic Description: Large evergreen trees, with fragrant White flowers & 3 valved pale Brown capsules fruits.

Medicinal Parts: Bark, Resin, Oil.

Chemical Constituents: Bark-Polyphenols as Epi-catechin, Levorototory, Isomers of Fisetinidol, dl-epicatechin, Afzelechin & Coumarine as Bergenin.

Resin - An essential oil contains □-Camphene, □-Camphor, □-Rinene, β-Rinene, Limonene, Chamazulene, Caryophyllene, Thymol.

***Medicinal Properties*:**

अजकर्णः कटुस्तिक्तः कषायोष्णो व्यपोहति ।
कफपाण्डुश्रुतिगदान् मेहकुष्ठविषव्रणान् ।। (भा. प्र.)

Ajakaṛṇaḥ katuṣtiktaḥ kāsayoṣṇo vyapohati
Kaphapāṇdusṛutigadāṇ mehakuṣthavisavṛaṇāṇ. (Bhā. Pṛa.)

Bark - Acrid, Alexeteric, Bitter, Depurative, Expectorant, Thermogenic, Tonic, Vulnerary.

Resin - Alexeteric, Bitter, Carminative, Expectorant, Tonic.

Fatty oil obtained from seeds - Antibacterial.

Medicinal Uses: Bark – Anaemia, Asthma, Cough, Dysentery, Leprosy, Skin eruptions, Ulcers, Wounds.

Resin - Amenorrhoea, Chronic Bronchtis, Diarrhoea, Dysmenorrhoea, Haemorrhoids, Hemicrania, Ringworm, Tubercular Glands.

Fatty oil obtained from seeds – Neuralgia, Rheumatism.

VENTILAGO MADERASPATANA Gaertn. (RHAMNACEAE)

Syn.: *V. calyculata* Tul.; *V. denticulata* Willd.

Regional & Other Names : *Bengali* – Raktapita; *English* – Red creeper; *Gujarati* - Ragatarohado; *Hindi* – Pitti; *Kannada* – Papudi; *Malayalam* – Vempata; *Marathi* –Kanvel; *Sanskrit* – Raktavalli; *Tamil* – Vempatam; *Telugu* – Suralatige.

Distribution: Throughout India.

Taxonomic Description: Large woody climbers, with Greenish flowers & Yellowish globular nuts.

Medicinal Parts: Bark.

Chemical Constituents: Stem bark - Friedelin.

Root bark - Trihydroxy-methyl anthranolmonomethyl Ether, Emodinmonomethyl ether.

Fruits, Leaves, Stems - Lupeol, β-Sitosterol & Glucoside.

Medicinal Properties:

वल्लीपर्यायशब्दस्तु सूर्यपर्यायपूर्वकः ।
भूते पतंगं तु यथा सूर्यवल्ली भवेदिति ।। (अ. म.)

Vaḷḷīpaṛyāyaśabdaṣtu sūṛyapaṛyāyapūṛvakaḥ
Bhrūte pataṇgaṃ tu yathā sūṛyavaḷḷī bhavediti. (A. Ma.)

Bark - Alexeteric, Astringent, Bitter, Carminative, Depurative, Digestive, Stimulant, Stomachic, Thermogenic, Tonic, Vulnerary.

Medicinal Uses: Bark - Colic, Dyspepsia, Erysipelas, Fever, Flatulence, General debility, Leprosy, Pruritus, Scabies, Skin diseases.

Ethno- Medicinal Uses: Root - Snake bite (Maheshwari *et al.,* 1986); Earache (Tarafder & Rai Chaudhuri, 1981).

Flowers - Earache (Tribedi *et al.*, 1982).

Stem - Eye diseases (Maheshwari & Singh, 1987); Eye pain (Udayan, *et al.,* 2006).

Bark - Rheumatism (Saxena *et al.,* 1981); Spermatorrhoea (Singh & Maheshwari, 1983); Sprain (Jain & De, 1966); Stomach ulcers (Pal & Jain, 1989) (Pal & Jain, 1989).

Fruits - Skin diseases (Sen *et al.,*1984).

Twig - Wounds. (Prasad et. al., 2003).

Root Bark - Stomach pain Fever, Skin diseases, Tonic (Kothari & Rao, 2003).

VIGNA PILOSA Baker. (FABACEAE)

Syn.: *Dolichos pilosus* Roxb.

Regional & Other Names : *Bengali* – Jhikrai; *Hindi* – Banmung; *Kannada* – Kaduhesaru; *Malayalam* – Kattupayar; *Sanskrit* – Mudgaparni; *Tamil* – Kattuppayar; *Telugu* – Adavipesalu.

Distribution: Throughout India.

Taxonomic Description: Slender hairy twiners, with Reddish or Bluish flowers & pendulous, linear, very long pointed velvety pods.

Medicinal Parts: Roots.

Medicinal Properties:

मुद्गपर्णी हिमा स्वादुर्वातरक्तविनाशिनी ।
पित्तदाहज्वरान् हन्ति कृमिघ्नी कफशुक्लकृत् ।। (ध. नि.)

Mudgapaṛṇī himā ṣvāduṛvātaraktavināśinī
Piṭṭadāhajvarāṇ haṇit ḳrmighnī kaphaśuklakṛṭ. (Dha. Ni.)

Roots - Aphrodisiac, Bitter, Cooling, Sweet.

Medicinal Uses: Roots - Burning sensation, Cough, Diarrhoea, Dyspepsia, Fever, Haemorrhoids.

VIGNA RADIATA (L.) Wilczek (FABACEAE)

Syn.: *Phaseolus radiatus* L. non Roxb. & Auct.

Regional & Other Names : *English* –Green gram; *Hindi* – Mung; *Kannada* – Hesaru; *Malayalam* – Cerupayar; *Sanskrit* – Mudgah; *Tamil* – Paccappayaru; *Telugu* – Pesalu.

Distribution: Throughout India, Cultivated.

Taxonomic Description: Twiners, with trifoliate leaves & thin, subcylindrical pods fruits.

Medicinal Parts: Seeds.

Medicinal Properties:

मुद्गः कषायो मधुरः कफपित्तास्रजिल्लघुः ।
ग्राही शीतः कटुः पाके चक्षुष्यो नातिवातलः ।। (रा. नि.)

Mudgaḥ kasāyo madhuraḥ kaphapittāsrajillaghuḥ
Grāhi sitaḥ katuḥ pāke caksusyo nativātalaḥ. (Ra. Ni.)

Seeds - Acrid, Aphrodisiac, Astringent, Constipating, Depurative, Digestive, Diuretic, Galactagogue, Styptic, Sweet.

Medicinal Uses: Seed- Asthma, Dropsy, Fever, General debility, Gonorrhoea, Leucoderma, Scabies, Skin diseases, Strangury.

Ethno- Medicinal Uses: Seed - Gout (Jadeja *et al.,* 2006b); To make strong & silky hairs (Jadeja *et al.,* 2006c).

VIGNA RADIATA (L.) Wilczek var. ***SUBLOBATA*** (Roxb.) Verdc. (FABACEAE)

Syn.: *Phaseolus sublobatus* Roxb.

Regional & Other Names : *English* – Wild black gram; *Hindi* – Banurd; *Kannada* – Kadu-Uddu; *Malayalam* – Kattulunnu; *Sanskrit* – Masaparni; *Tamil* – Kattu-Uluntu; *Telugu* – Adavi-Uddulu.

Distribution: Throughout India.

Taxonomic Description: Twiners, with trifoliate leaves & thin, subcylindrical pods fruits.

Medicinal Parts: Roots.

Medicinal Properties:

माषकर्णी हिमा तिक्ता रूक्षा शुक्लवलासकृत् ।
मधुरा ग्राहिणी शोफवातपित्तज्वरास्रजित् ।। (भा. प्र.)

Māsparṇī himā tiktā rūksā śuklavalāsakrt
Mādhurā gṛāhiṇī śophavātapiṭṭajvarāsṛajiṭ. (Bhā. Pṛa.)

Roots-Anti-Inflammatory, Aphrodisiac, Bitter, Constipating, Depurative, Digestive, Emollient, Febrifuge, Refrigerant, Sweet.

Medicinal Uses: Roots - Burning sensation, Colic, Cough, Diarrrhoea, Dyspepsia, Faltulence, Fever, General debility, Haemorrhages, Inflammation, Leprosy, Seminal weakness, Skin diseases.

VIGNA UNGUICULATA (L.) Walp. ssp. ***CYLINDRICA*** (L.) Eselt. (FABACEAE)

Syn.: *V. sinensis* (L.) Hassk.; *V. catjang* (Burm. f.) Walp.

Regional & Other Names : *Bengali* – Barabati; *English* – Cow pea; *Gujarati* - Chola; *Hindi* – Sonta; *Kannada* – Alasandi; *Malayalam* – Payar; *Marathi*–Chavali; *Sanskrit*–Chavala; *Tamil*–Karamani; *Telugu* – Boberulu.

Distribution: Throughout India.

Taxonomic Description : A suberect or twining annual. Leaflets 3, entire or slightly lobed. Flowers Yellow or Reddish. Pods linear, subtorulose.

Medicinal Parts: Seeds.

Chemical Constituents: Seed contains-Carotene, Thiamine, Riboflavin, Niacin, Folic acid, Vitamin-C, Phytohemagglutinin, □-Cedrene, Hexylacetate, Limonene, Nonanal, β-Ocinene, □-Pinene, β-Pinene.

Medicinal Properties:

राजभाषः सरो रूच्यः कफपित्तास्रशुक्लनुत् ।
सुस्वादुः वातलो रूक्षः कषायो विशदो गुरूः ।। (ध. नि.)

Rājamāsaḥ saro rūcyaḥ kaphapiṭṭāsṛaśuklanuṭ
Suṣvāduḥ vātalo rūksaḥ kasāyo viśado gūruḥ. (Dha. Ni.)

Seeds - Anaphrodisiac, Anthelmintic, Appetiser, Astringent, Diuretic, Laxative, Liver tonic, Sweet.

Medicinal Uses: Seeds - Agalactia, Anorexia, Constipation, General debility, Helminthiasis, Jaundice, Strangury.

Ethno- Medicinal Uses: Px - Small pox (Jain & Tarafder, 1970).

Seed - Diuretic, Anthelmintic (Basak, 2006).

WEDELIA CHINENSIS (Osbeck.) Merrill. (ASTERACEAE)

Syn.: *W. calendulacea* Less.

Regional & Other Names : *Bengali* – Bhringaraj; *Gujarati* - Bhangaro; *Hindi* – Bhangra; *Kannada* – Kaesarji; *Malayalam* – Aswagandhi; *Marathi* – Pivalamaka; *Sanskrit*-Bhringaraja; *Tamil* – Tatalaikaintageri.

Distribution: Throughout India.

Taxonomic Description: Procumbent perennial herbs, with Yellow flowers & achenes fruits.

Medicinal Parts: Whole plant.

Chemical Constituents: Plant contains Coumestan, Norwedelic acid, Norwedelotactone, Tri-O-Methylwedelolactone & β-amyrin.

Medicinal Properties: Whole plant - Astringent, Bitter, Acrid, Thermogenic, Anti-Inflammatory, Vulnerary, Cardiotonic, Anthelmintic, Diuretic, Aphrodisiac, Sudorific, Febrifuge.

Medicinal Uses: Whole plant - Anaemia, Colic, Dyspepsia, Elephanthiasis, Fever, Helminthiasis, Inflammation, Otalgia, Strangury, Seminal weakness Ulcers, Wounds.

Ethno-Medicinal Uses: Whole plant - Uterine haemorrhages, Menorrhagia (Gogoi *et al.,* 2003).

Leaf - Jaundice (Ravikumar & Sankar, 2003).

Stem / leaf - Hair growth (Das *et al.,* 2006).

ZANONIA INDICA L. (CUCURBIACEAE)

Regional & Other Names : *English* – Bandolier fruits; *Gujarati* - Parpoti; *Hindi* – Chirpoti; *Malayalam* – Penarvalli; *Marathi* – Chirabuti; *Sanskrit* – Kuntali.

Distribution: Throughout India.

Taxonomic Description: Stout climbing shrubs, with Greenish, Yellow flowers & pale Yellowish or Brown cylindrical capsules.

Medicinal Parts: Leaves, Fruits.

Medicinal Properties:

चिर्पोटा हिमा रूक्षा भेदिनी श्वासकासजित् ।। (शा. नि.)

Chiṛpotā himā rūksā bhedinī ṣvāsakāsajiṭ. (Śā. Ni.)

Leaves - Anti-Inflammatory, Antispasmodic.

Fruits-, Antiseptic, Cooling, Expectorant, Purgative,

Medicinal Uses: Leaf - Inflammation, Spasmodic affections.

Fruits- Constipation, Cough, Asthma, Cuts, Wounds, Ulcers.

ZIZIPHUS NUMMULARIA (Burm. F.) Wt. & Arn. (RHAMNACEAE)

Syn.: *Z. rotundifolia* Lam.; *Rhamnus nummularia* Burm f.

Regional & Other Names : *Bengali* – Kantaikolli; *English* – Wild jujube; *Gujarati* - Chanyabor; *Hindi* – Jharberi; *Marathi* – Janglar; *Sanskrit* –Kshudrabadar; *Tamil* – Korgodi; *Telugu* – Parimpalu.

Distribution: Throughout India.

Taxonomic Description: Much-branched, prickly shrubs, with zigzag branches & orbicular-ovate leaves tomentose above. Flowers greenish-yellow, in axillary cymes. Drupes globose, red when ripe.

Medicinal Parts: Leaves, Fruits.

Chemical Constituents: Twig bark contains Tannin.

Root bark contains Cyclopeptide alkaloids Nummularine A, B, C; Mucronine-D & Amphibine-H.

Medicinal Properties:

वातघ्नं कफकृत्तद्वत् कोलमुष्णं सरं गुरू ।
तद्वत् कर्कन्धु पक्कानि स्वादु स्निग्धरसानि च ।। (कै. नि.)

Vātaghnaṃ kaphakruṭ kolmusnaṃ saraṃ guru
Taḍvaṭ karkaṇdhu pakkāni svādu snigdhrasāni ca. (Kai. Ni.)

Fruits - Appetiser, Astringent, Cooling, Sour, Stomachic, Sweet.

Medicinal Uses: Leaves - Cold, coughs, Scabies, Skin diseases.

Fruits - Bilious affections.

Ethno- Medicinal Uses: Leaf - Boils (Rajwar, 1983); To arrest falling hairs (Ghosh, 2006).

Fruits - Cold (Shekhawat & Sunil Anand, 1984); Indigetion (Rajwar, 1983).

Root Bark - Pyorrhoea (Kumar & Chauhan, 2006); Dysentery (Rajwar, 1983). Inflammation in gums (Rajwar, 1983).

Root- Diarrhoea (Audichya *et al.,* 1983); tonic (Shah *et al.,* 1983); Fever (Singh & Kumar, 2003).

ZIZIPHUS OENOPLIA (L.) Mill. (RHAMNACEAE)

Regional & Other Names : *Bengali* – Siakul; *English* – Jackal Jujube; *Hindi* – Makai; *Kannada* –Challe; *Malayalam* –Kottavalli; *Marathi* –Makor; *Sanskrit* – Srigalakoli; *Tamil* – Paragi; *Telugu* –Banka.

Distribution: Throughout India.

Taxonomic Description: A straggling or climbing shrub. Prickles paired one straight the other recurved. Leaves ovate, pointed; base oblique. Flowers in short axillary, dichotomous cymes. Drupes globose or obovoid, Black, shining.

Medicinal Parts: Roots.

Chemical Constituents: Bark contain tannin.

Stem & Root bark contain Cyclopeptide alkaloids Zizyphine-A & Zizyphine-B.

Medicinal Properties:

लघुबदरं मधुराम्लं पक्वं कफवातनाशनं रूच्यम् ।
स्निग्धं तु जन्तुकारकमीषत् पित्तार्त्तिदाहशोषघ्नम् ।। (रा. नि.)

Laghubadaraṃ madhuraṃlaṃ pakvaṃ kaphavatanasanaṃ rucyaṃ
Snigdhaṃ tu jaṇtukarakamisaṭ piṭtarttdiahasosaghnaṃ. (Ra. Ni.)

Roots - Anthelmintic, Antiseptic, Astringent, Bitter, Digestive.

Medicinal Uses: Roots - Ascaris Infection, Hyperacidity, Stomachalgia, Wounds.

Ethno-Medicinal Uses: Root - Abdominal pain (Tarafder, 1983) (Tarafder, 1983); Swelling (Shah *et al.,* 1983); Wounds (Janardhanan, 1963).

Stem - Dysentery (Rajwar, 1983).

Fruits - Blood purification, Indigestion (Rajwar, 1983); Stomach diseases (Sinha *et al.,* 2006).

Leaf - Snake bite (Murthy *et al.,* 1986); Conjunctivitis (Rajwar, 1983).

Bark - Indigestion (Shah *et al.,* 1981); Eczema (Satapathy, 1992).

GLOSSARY OF MEDICAL TERMS

Abortifacient	an agent that induces abortion
Acrid	biting, pungent
Adenitis	inflammation of a gland or group of glands (or glandlike structures)
Agalactia	absence or failure of secretion of milk
Alexipharmic	antidote to poison
Alexiteric	protective to infectious diseases
Alopecia	loss of hair: a malady in which the hair falls from one or more circumscribed round or oval areas, leaving the skin smooth and white
Alterative	causing a favourable change in the disordered functions of the body or metabolism
Amenorrhoea	failure of menstruation
Amentia	an arrest of the development of the mind from birth to early age
Anaemia	lack of enough blood causing paleness
Analgesic	an anodyne
Anaphrodisiac	having the power to lessen or inhibit sexual feeling
Anodyne	a medicine that allays pain
Anorexia	a condition of having lost the appetite for food
Anthelmintic	destroying or expelling worms
Antibacterial	active against bacteria
Antidote	an agent which neutralises or opposes the action of a poison
Anti-inflammatory	acting against inflammation
Anti-leprotic	any agent which prevents or cures leprosy
Antilithic	an agent which prevents the formation of calculi
Anti-periodic	preventing the regular recurrence of a disease
Antipyretic	counteracting fever
Antiscorbutic	acting against scurvy

Anti-septic	a chemical sterilising substance to kill or control pathogenic microbes
Antispasmodic	opposing Spasms
Anuria	failure of the kidneys to produce urine
Aperient	A laxative
Aphrodisiac	drug which stimulates sexual desire
Areata	ocurring in circumscribed areas or patches
Arthritis	inflammation of joints
Ascities	an abnormal accumulation of a fluid in the peritoneal cavity
Astringent	a drug which contracts the organic tissues & canals of the body, & therby checking or deminishing bleeding of excessive discharges
Azoospermia	the comlete absence of sperm from the seminal fluid
Baldness	absence of hair from areas where it normally grows
Beriberi	a deficiency disease caused by imbalance of carbohydrate and vitamin B
Biliousness	a symptom complex comprising nausea
Boil	tender inflamed area of the skin containing pus
Bronchitis	inflammation of the bronchi
Bronchodilator	an agent that causes widening of the air passages by relaxing bronchial smooth muscle
Bubo	an inflammatory swelling of a lymph gland
Calculi	a hard pebble-like mass formed within the body, particularly in the gall bladder or anywhere in the urinary tract
Cancer	the abnormal and uncontrolled division of cells that then invade and destroy the surrounding tissues
Carbuncle	an infection of the skin and subcutaneous tissue by Staphylococcus aureus
Carminative	drug curing flatulence
Catarrh	inflammation of a mucous membrane
Cathartic	having the power of cleansing the bowels- purgative
Cephalagia	headache
Cephalic	a remedy for disorders of the head
Cholera	a severe infectious epidemic disease due to Vibrio cholerae
Chronic bronchitis	the patient coughs up excessive mucus secreted by enlarged bronchial mucous glands
Conjunctivitis	inflammation of conjunctiva
Colic	a severe spasmodic griping pain

Consumption	pulmonary tuberculosis
Contraceptive	drug which prevents unwanted pregnancy
Convulsion	a violent involuntry contraction of the skeletal musculature
Cystitis	inflammation of a bladder, especially the urinary bladder
Dandruff	dead scarf-skin seperating in small scales and entangled in the hair
Deafness	partial or total loss of hearing in one or both ears
Demulcent	soothing
Depurative	an agent that purifies the blood
Diabetes	here it refers to diabetes mellitus: a disorder of carbohydrate metabolism in which sugars in the body are not oxidized to produce energy due to lack of the pancreatic hormone-insulin. The accumulation of sugar leads to its appearance in the blood
Diaphoretic	a drug which induces perspiration
Diarrhoea	frequent bowel evacuation or the passage of abnormally soft or liquid faeces
Diptheria	a specific infectious disease caused by virulent strains of a bacillus
Diuretic	promoting the discharge of urine
Dropsy	an excessive accumulation of a clear or watery fluid in any of the tissues or cavities of the body
Dysentery	an infection of the intestinal tract causing severe diarrhoea with blood and mucus
Dysmenorrhoea	painful menstruation
Dysopia	defective vision
Dyspepsia	indigestion
Dysuria	difficulty or pain while passing urine
Earache	pain in the ear
Eczema	a noncontagious inflammatory disease of the skin with much itching and burning
Elephantiasis	gross lymphatic oedema of the limbs leading to hypertrophy
Emetic	causing vomiting
Emollient	softening
Epilepsy	an affection of the nervous system resulting from excessive or disordered discharge of cerebral neurons
Eruptions	the outbreak of a rash
Erysipelsa	an inflammatory disease generally on the face marked by a bright redness of the skin
Expectorant	aiding the secretion of the mucous membrane of the air passages & the removal of fluid by spitting
Fatigue	mental or physical tiredness, following prolonged or intense activity

Febrifuge	that which drives off fever
Fever	a rise in body temperature above the normal
Fistula	an open channel from the anus or rectum to the skin near the anus
Flatulence	distention of stomach
Foetid	bad-smelling, fetid
Gleet	chronic discharge of thin mucus from the vagina
Goitre	enlargement of the thyroid gland
Gonorrhoea	an inflammatory disease of the urinogenital passagaes characterised by pain and discharge
Gout	a disease of purine metabolism characterised by attacks of arthritis with an associated raised level of serum uric acid
Haematemesis	vomitting of blood
Haematinic	any substance which is required for the production of RBCs & its constituents
Haemoptysis	spitting of blood
Haemorrhage	the escape of blood from a ruptured blood vessel, externally or internally
Haemorrhagic diseases	associated with or resulting from blood loss
Haemostatic	styptic
Hallucination	a false perception of something that is not really there
Headache	pain felt deep within the skull
Helminthiasis	morbid state due to infestation with worms
Helminthic	Anthelmintic, vermifugal
Hemicrania	headache confined to one side
Hepatomegaly	enlargement of the liver
Hiccough	abrupt involuntary lowering of the diaphragm and closure of the sound-producing folds at the upper end of the trachea, producing a characteristic sound as the breath is drawn in
Hydrocele	a circumscribed collection of fluid in the tunica vaginalis testis
Hydrophobia	exaggerated fear of water as in rabies
Hypertension	high arterial blood pressure
Hysteria	a neurotic disorder with varying symptoms
Impotency	inability to perform the sexual act due to failure of the reflex mechanism
Inflammation	the body's response to injury, which may be acute or chronic
Insecticidal	any agent which kills or destroys insects

Insomnia	the condition a being unable to sleep
Intermittient fever	a fever that rises, subsides, then returns again
Itcing	discomfort or irritation of the skin, prompting the sufferer to scratch or rub the affected area
Jaundice	a yellowing of the skin or whites of the eyes, indicating excess bilirubin(a bile pigment) in the blood
Laxative	a drug used to stimulate or increase the frequency of bowel evacuation, or to encourage the passage of a softer or bulkier stool
Leprosy	a chronic disease, caused by the bacterium Mycobacterium leprae, that affects the skin, mucous membranes, and nerves
Leucoderma	loss of pigments in areas of the skin, resulting in the appearance of white patches or bands
Leucoma	white opacity occurring in the cornea
Leucorrhoea	whitish or yellowish discharge of mucus from the vaginal opening
Lumbago	pain in lower back
Mammillitis	inflammation of the nipple
Menorrhagia	excessive or prolonged menstruation
Migrain	condition resulting from spasm and subsequent overdilatation of certain arteries in the brain, which causes a recurrent throbbing headache that characteristically affects one side of the head
Miscarriages	the expulsion or removal of an embryo or fetus from the uterus at a stage of pregnancy when it is incapable of independent survival
Mumps	an acute infectious disease caused by a virus
Nasitis	inflammation of the nose
Nightblindness	the inability to see in dim light or at night
Obesity	an excessive accumulation of fat in the body
Oedema	excessive accumulation of fluid in the body tissues
Oleaginous	oily
Ophthalmia	a term usually applied to conjunctivitis
Ophthalmopathy	any disease of the eye
Otalgia	pain in the ear
Otorrhoea	any discharge from the ear, commonly a purulent discharge in chronic middle ear infection
Paralysis	a muscle weakness that varies in its extent, its severity, and the degree of spasticity or flaccidity according to the nature of the underlying disease and its distribution in the brain, spinal cord, peripheral nerves, or muscles
Pectoral diseases	effective in diseases of the chest

Peptic ulcer	a breach in the lining(mucosa) of the digestive tract produced by digestion of the mucosa by pepsin and acid
Piles	enlargement of the normal spongy blood-filled cushions in the wall of the anus, usually a consequence of prolonged constipation or ocassionally, diarrhoea
Pneumonia	a general disease in which the essential lesion is an inflammation of the spongy tissue of the lung with consolidation of the alveolar exudate
Pneumonosis	any lung disorder
Polyuria	the production of large volumes of urine, which is dilute and of a pale colour
Procreant	the drug which begets
Prophylactic	pertaining to the prevention of the development of a disease
Prurigo	an eruption of the skin causing severe itching
Pruritus	itching
Pyorrhoea	disease of the tissues that support and attach the teeth-the gums, periodontal membrane, and alveolar bone
Refrigerant	cooling
Remittent fever	a remittent fever is one in which body temperature flactuates but does not return to normal
Renal calculi	calculi relating to kidney
Resolvent	causing resolution of a tumour or swelling
Retention of urine	inability to pass urine, which is retained in the bladder
Revulsive	causing revulsion in drawing away of blood from a pathological area to another area
Rheumatalgia	rheumatic pain
Rheumatism	any disorder in which aches and pains affect the musclas and joints
Rheumatoid arthritis	it is the second most common form of arthritis. It typically involves the joints of the fingers, wrists, feet and ankles with later involvement of the hips, keens, shoulders and neck
Rhinitis	inflammation of the nasal mucous membrane
Ringworm	fungus infection of the skin, the scalp, or the nails
Rubefacient	reddening of the skin
Scabies	sarcoptic infestation of the human skin particularly a contagious skin disease caused by invasion of the epidermis
Sciatica	pain felt down the back and outer side of the thigh, leg, and foot
Scrofula-	tuberculous cervical adenitis with or without ulceration
Scurvy	a deficiency disease due to lack of vitamin C

Sinusitis	inflammation of one or more of the mucous- membrane-lined air spaces in the facial bones that communicate with the nose
Slipdisc	a colloquial term for a prolapsed intervertebral disc
Small Pox	an acute infectious virus disease causing high fever and a rash that scars the skin
Somatalgia	pain in the body
Soporific	inducing sleep
Sore throat	pain at the back of the mouth, commonly due to bacterial or viral infection of the tonsils
Spasmolytic	a drug that relieves spasm of smooth muscle
Splenopathy	any disease of the spleen
Spermatorrhoea	the involuntary discharge of semen without orgasm
Stomachache	pain in the stomach
Stomachic	a medicine that increases the functional activity of the stomach
Stomatitis	generalised inflammation of the oral mucosa
Strangury	severe pain in the urethra referred from the base of the bladder and associated with an intense desire to pass urine
Styptic	having the power to arrest bleeding
Sudorific	an agent which induces sweating
Sun stroke	raised body temperature (pyrexia), absence of sweating, and eventual loss of consciousness due to failure or exhaustion of the temperature-regulating mechanism of the body
Syphilis	contagious venereal disease
Tetanus	an infective disease due to toxin of Clostridium tetani
Tuberculosis	an infectious disease caused by the bacillus Mycobacterium tuberculosis and characterized by the formation of nodular lesions in the tissues
Tumour	any abnormal swelling in or on a part of the body
Typhoid	an infection of the digestive system by the bacterium Salmonella typhi, causing general weakness, high fever, a rash of red spots on the chest and abdomen
Ulcers	a break in the skin extending to all its layers, or a break in the mucous membrane lining the alimentary tract, that fails to heal and is often accompained by inflammation
Urethrorrhoea	abnormal discharges from the urethra
Urticaria	an itchy rash resulting from the release of histamine by mast cells
Uteralgia	pain in the uterus
Veneral diseases	any disease transmitted by sexual intercourse

Verminosis	infestation with vermin
Vertigo	dizziness
Vesical calculi	stone in the urinary bladder
Vomiting	the reflex action of ejecting the contents of the stomach through the mouth
Vulnerary	an agent that promotes healing of wounds
Warts	a circumscribed cutaneous excrescence
Whooping cough	an acute infectious disease, primarily of infants, caused by the bacterium Bordetella pertusis, which affects the respiratory tract
Wounds	a break in the structure of an organ or tissue caused by an external agent

BIBLIOGRAPHY

Abraham, Z. 1981. Ethnobotany of the Todas, the Kotas and the lrulas of the Nilgiris, 308 – 320. In S.K. Jain (Ed.) *Glimpses of Indian Ethnobotany.*

Aditya, P., A. Kumar & D. K. Yadav. 2006. Herbal renedies for Dental care system in Gaya district, Bihar. *J. Econ. Taxon. Bot.* 30(Suppl.): 167-170.

Agarwal, V. S. & Arya, M. 2005. *Alternanthera sessilis* DC. (Amaranthaceae): A new source for the treatment of rickets and marasmus. *Ethnobotany.* 17:200-201.

Ahluwalia, K.S. 1968. Medicinal plants of Kerala-VI. *Nagarjun* 11: 392-382.

Ahmad, A. 1995. Drugs of plant origin as used by certain tribes of eastern (Purvanchal) U. P. (Part -IV). *Int J Mendel 12 (1-4),* 12 – 14.

Ahmad, A. 1996. Drugs of plant origin as used by certain tribes of eastern (Purvanchal) U. P. (Part -V). *Indian J Appl Pure Biol:* 11: 41-52.

Aminduddin, Girach, R. D. & Khan, 1993 Treatment of Malaria through herbal drugs from Orissa. *Fitoterapia* 64: 545 – 548.

Amit tomar & H. Singh. 2006. Exotic medicinal plants from Baghpat district, Uttar Pradesh *J. NTFPs* 13(4): 273-280.

Apparanantham, T. & V. Chelladurai, 1986. Glimpses on folk medicines of Dharmapuri forest division, Tamil Nadu, *Anc. Sci. Life.* 5: 182 – 185.

Ara, Shoukat & A. R. Naqshi. 1992. Ethnobotanical studies in Gurais valley- Kashmir. *J. Econ. Taxon. Bot. Addl. Ser.* **10:** 185-191.

Arinathan, V., V. R. Mohan & A. John De Britto. 2003. Ethnomedicinal survey among Palliyar tribals of Srivilliputhur Grizzled giant squirrel wildlife sanctuary, Tamil Nadu. 707-710. In: Singh *et al.* (ed.). Ethnobotany and Medicinal Plants of India and Nepal. Vol-II. Scientific Publishers (India), Jodhpur.

Arinathan, V., A. Amuthavalli & M. Malaiarasi. 2006. Studies of medicinal plants in Thoothukudi, Tamilnadu. *J. Econ. Taxon. Bot.* 30(Suppl.): 353-357.

Arinathan, V., J. Joseph Visuvasam & R. G. Muthu. 2006. Survey of medicinal plants in Kulathur, Tamilnadu. *J. Econ. Taxon. Bot.* 30(Suppl.): 370-374.

Arya, K. R., Pande, P. C. & Prakash, V. 1999. Ethnobotanical study on tribal areas of Almora district-II. *Ethnobotany.* 11:100-104.

Arya, K. R. & Ved Prakash. 2003. Ethnomedicinal study of a remote tribal area of Almora district: A survey report- Part-I. 247-252 In: Maheshwari, J. K. (ed.). Ethnobotany and Medicinal Plants of Indian Subcontinent. Scientific Publishers (India), Jodhpur.

Arya, K. R. & S. C. Agarwal. 2006. Conservation of threatened medicinal plants through cultivation in Uttaranchal State. *Ethnobotany*. 18:77-86.

Audichya, K. C., K. V. Billore, T. H. Joseph & D. D. Chaturvedi, 1983. Role of indigenous folk remedies for certain acute illnesses in primary health care. *Nagarjun*.26: 199 – 201.

Bajpai, A. Ojha, J.K. & Sant H.R. 1995. Medicobotany of the Varanasi district, U.P. India. *Int. J. Pharmacog.* 33: 172-176.

Balaji Rao, N.S. Rajasekhar, D., Chengal Raju, D. & Nagaraju, N. 1996. Ethno-medicinal notes on some plants of Tirumala hills for dental disorders. *Ethnobotany* 8: 88-91.

Ballabh, Basant & O. P. Chaurasia. 2006. Ethnobotanical studies on Boto tribe in Ladakh. *Ethnobotany* 18: 87-95.

Balodi, B. & Singh, D.K. 1997. Medico-ethnobotany of Ladakh. *Ann For* 5: 189-197.

Banerjee, A.K. & Banerjee, I. 1986. A Survey of the medicinal plants in Shevaroy hills. *J. Econ. Taxon. Bot.* 8: 271-290.

Banerjee, A. 2003. Ethnobotany of some trees in the Santhal villages of Birbhum, West Bengal. 531-534. In: Maheshwari, J. K. (ed.). Ethnobotany and Medicinal Plants of Indian Subcontinent. Scientific Publishers (India), Jodhpur.

Barua, K. N., I. C. Barua & Mrigen Das. 2003. Ethnobotany of Rajbanshis of Assam. 609-614. In: Maheshwari, J.K. (ed.). Ethnobotany and Medicinal Plants of Indian Subcontinent. Scientific Publishers (India), Jodhpur.

Baruah, P. & G.C. Sarma. 1984. Studies on the medicinal uses of plants by the Boro tribals of Assam II. *J. Econ. Taxon. Bot.* 5: 599 – 604.

Baruah, Parukutty & G. C. Sarma, 1987. Studies on the medicinal uses of plants by the Northeast tribes – III. *J.Econ. Tax. Bot.* 11: 71– 76.

Baruah, M. K., P. D. Choudhury & G. C. Sarma. 2006. Ethnomedicinal plants used by Khasi tribe of Cachar district, Assam. *J. Econ. Taxon. Bot.* 30(Suppl.): 110-114.

Basak, S. K. 1997. Medicinal plants of Bankura (W. B.) and their uses. *J. Nat Bot Soc.* 51: 61-68.

Basak, S. K. 2006. Medicinal plants of Bankura (West Bengal) and their uses. *J. Econ. Taxon. Bot.* 30(Suppl.): 217-224.

Basu, R. 2003. Ethnomedicinal information of yellow flowered 'Palas' and 'Silk cotton' in Bankura District of West Bengal. 580-581. In: Singh *et al.* (ed.). Ethnobotany and Medicinal Plants of India and Nepal. Vol-II. Scientific Publishers (India), Jodhpur.

Bedi, S. J. 1978. Ethnobotany of the Ratan Mahal Hills, Gujarat, India, *Econ. Bot.*32: 278 – 284.

Behura, S. & N. C. Rout. 2003. Diversity of Zingiberaceae plant from Orissa. 546-550. In: Singh *et al.* (ed.). Ethnobotany and Medicinal Plants of India and Nepal. Vol-II. Scientific Publishers (India), Jodhpur.

Beigh, S. Y., Irshad A. Nawchoo & Muhammad Iqbal. 2003. Plants in traditional medicine of Kashmir Himalaya. 99-104. In: Singh *et al.* (ed.). Ethnobotany and Medicinal Plants of India and Nepal. Vol-I. Scientific Publishers (India), Jodhpur.

Bennet, S.S.R. 1985. Ethnobotanical studies in West Sikkim. *J. Econ. Taxon. Bot.*7: 317 – 321.

Bhalla, N.P., T.R. Sahu, G.P. Mishra & R.N. Dakwale. 1982. Traditional Plant medicines of Sagar Distt. Madhya Pradesh, India. *J. Econ. Taxon. Bot.*3: 23 – 32.

Bhalla, S., J.R. Patel & N.P. Bhalla. 1992. Ethnomedicinal herbal Legumes of Bundelkhand region, M.P. *J. Econ. Taxon. Bot. Addl. Ser.* **10:** 105-109.

Bhandary, M.J. & K.R. Chandrashekar. 2002. Glimpses of ethnic herbal medicine of Coastal Karnataka. *Ethnobotany.* **14:** 1-12.

Bhatt, R.P. & S.D. Sabnis, 1987. Contribution to the Ethnobotany of Khedbrahma region of north Gujarat. *J. Econ. Taxon. Bot.*9: 139 – 145.

Bhatt, D. C., N. K. Patel, K. D. Mitaliya & H. M. Ant. 2003. Herbal magic by contact therapy among tribals and rurals of Gujarat. 205-207. In: Singh *et al.* (ed.). Ethnobotany and Medicinal Plants of India and Nepal. Vol-I. Scientific Publishers (India), Jodhpur.

Bhatt, D. C., N. K. Patel, K. D. Mitaliya & H. M. Ant. 2003. Herbal magic by contact therapy among tribals and rurals of Gujrat. 205-207. In: Singh *et al.* (ed.). Ethnobotany and Medicinal Plants of India and Nepal. Vol-I. Scientific Publishers (India), Jodhpur.

Bhatt, D. C., K. D. Mitaliya & A. V. Babaria. 2003. Observations on some ethnobotanical plants of Malnath Hillocks of Gogha Tehsil, district Bhavnagar, Gujarat State. 217-220. In: Singh *et al.* (ed.). Ethnobotany and Medicinal Plants of India and Nepal. Vol-I. Scientific Publishers (India), Jodhpur.

Bhatt, D.C., S.K. Mehta, M.A. Nurani & U.S. Baxi. 2003. Studies on some ethnobotanical plants of Barda Hills from Saurashtra. 221-227. In: Singh *et al.* (ed.). Ethnobotany and Medicinal Plants of India and Nepal. Vol-I. Scientific Publishers (India), Jodhpur.

Bhatt, D.C., B. A. Jadeja, N. K. Odedra & U. S. Baxi. 2003. Enumeration of wild plants used as anti-diabetic in Barda hills of Gujarat. 897-903. In: Singh *et al.* (ed.). Ethnobotany and Medicinal Plants of India and Nepal. Vol-II. Scientific Publishers (India), Jodhpur.

Bhatt, D. C., A. M. Patel, P. P. Parekh & R. P. Parmar. 2003. Plants used for sexual debility in Central Gujarat. 943-946. In: Singh *et al.* (ed.). Ethnobotany and Medicinal Plants of India and Nepal. Vol-II. Scientific Publishers (India), Jodhpur.

Bhattacharjee, S., K.C. Tiwari, R. Majuumdar & A.K. Mishra. 1980. Folklore medicine from district Kamrup (Assam). *Bull. Medico – ethnobot. Res.*1: 447 – 460.

Bhogaonkar, P. V. & V. N. Kadam. 2006. Ethnomedicobotany of some local reproductive health practices in Umarkhed of Yavatmal district in Maharashtra, India. *Ethnobotany.* 18:107-113.

Bhuyan, L. R. 2003. Some plants used as medicine by the Nishi tribe of Arunachal Pradesh: A preliminary study. 447-450. In: Singh *et al.* (ed.). Ethnobotany and Medicinal Plants of India and Nepal. Vol-I. Scientific Publishers (India), Jodhpur.

Billore, K.V. & K.C. Audichya. 1978. Some oral contraceptives; family planning tribal way, *J. Res. Indian Med. Yoga & Homeo.*13: 104.

Billore, K.V., Joseph, T.G. & Dave, S.K. 1998. Interesting folk remedies by the 'Lok, Vaidyas' of Rajasthan for 'Swas Roga'. *Ethnobotany.*10: 42 – 45.

Binu, S., A. E. Shanavaskhan, E. S. Santhoshkumar & P. Pushpangadan. 2003. Plants used as Medicine by the Irulas of Palghat district, Kerala, India. 808-814. In: Singh *et al.* (ed.). Ethnobotany and Medicinal Plants of India and Nepal. Vol-II. Scientific Publishers (India), Jodhpur.

Bondya, S. L., K. K. Khanna & K. P. Singh. 2006. Ethnomedicinal uses of leafy vegetables from the tribal folklore of Achanakmar- Amarkantak biosphere reserve (Madhya Pradesh & Chhattisgarh). *Ethnobotany.* 18:145-148.

Bora, P. J. 2003. A study on Ethnomedicinal uses of plants among the Bodo tribe of Sonitpur district, Assam. 604-608. In: Maheshwari, J.K. (ed.). Ethnobotany and Medicinal Plants of Indian Subcontinent. Scientific Publishers (India), Jodhpur.

Borthakur, S.K. 1976. Less known medicinal uses of plants among the tribals of Mikir Hills. *Bull. Bot. Surv. India.*18: 166 -171.

Borthakur, S.K. 1993. Native Phytotherapy for child and women diseases from Assam. *Ethnobotany* 75: 87-89.

Borthakur, S.K. & Goswami, N. 1995. Herbal remedies from Dimoria of kamrup district Assam in Northeastern India. *Fitoterapia* 66: 333-340.

Borthakur, S.K. 1996. Postnatal care of women in traditional system in Assam. *Ethnobotany.* 8: 51-55.

Borthakur, S.K. Nath, K. & Gogoi, P. 1996. Herbal remedies of the Nepalese of Assam, *fitoerapia.*67: 231 – 237.

Caius, J. F. 1998. The Medicinal and poisonous plants of India, scientific publishers, Jodhpur.

Chaghtai, S.A. & Javed Ahmad. 1978. An ethnobotanical study of Baghrendah (*Jatropha curcas* L.) *J. Res. Indian Med. Yoga & Homeo.*13: 134 – 136.

Chandra, K. & Pandey, B. N. 1985. Medicinal plants of santhal paragna district Dumka (Bihar) Part-I. Sachitra Ayurved 38, 307-314.

Chandra, K., Pandey, B.N. Sudha, G.N. & Pandey, P. 1989. Medicinal plants wealth of the Rajgir Bihar. *Bull. Med. Ethnobot. Res.* 1: 124-161.

Chandra, K. 1995. An Ethnobotanical study on some medicinal plants of District Palamau (Bihar). *Sachitra Ayurved.*48: 311 – 314.

Chatterjee, S. & A. Bhatterjee. 2006. Some Ethnomedicinal plants of North 24 Paragana, West Bengal. *J. Econ. Taxon. Bot.* 30(Suppl.): 281-282.

Chaudhuri, Rai H.N. & D.C. Pal. 1975. Ethnobotany of 'Vallatak'. *J. Sci.* Culb **29**:106-110.

Chaudhuri, Rai H.N., D.C. Pal & Tarafder. 1975. Less known uses of plants from the tribal areas of Orissa. *Bull. Bot. Surv India.* **17:** 132 – 136.

Chaudhuri, Rai H.N. & D.C. Pal. 1976. Preliminary observations on Ethnobotany of Midnapur district West Bengal. *India. Mus. Bull.* **11**: 51 – 53.

Chaudhuri, Rai H.N., H.A. Molla, D.C. Pal & B. Roy. 1982. Plants used in traditional medicine by some tribals of Jalpaiguri district, West Bengal. *Bull. Bot. Surv. India.* 24: 87-90.

Chaudhuri, Rai, H.N., D.C. Pal, N.C. Saha & B. Roy. 1985. Some wild edible plants in Calcutta markets. *Man & Life.*11: 45-58.

Chaudhury, D. & B. Neogi. 2003. Ethnobotany of Khasi and Chakma tribes of North East India. 583-589. In: Maheshwari, J. K. (ed.). Ethnobotany and Medicinal Plants of Indian Subcontinent. Scientific Publishers (India), Jodhpur.

Chauhan, V. & Chauhan, N.S. 1988. Ethno-botany of Trans-Giri area of Sirmour district of Himachal Pradesh, *Bull Med Ethnobot Res.* 9: 106-122.

Chaurasia, O. P., B. Singh & S. K. Sareen. 2003. Ethno-medico-botanical survey of Nubra Valley. 167-172. In: Maheshwari, J. K. (ed.). Ethnobotany and Medicinal Plants of Indian Subcontinent. Scientific Publishers (India), Jodhpur.

Chaurasia, O. P., B. Singh & D. P. Attrey. 2003. Ethnobotany of Changthang valley-Ladakh Himalaya. 108-113. In: Singh *et al.* (ed.). Ethnobotany and Medicinal Plants of India and Nepal. Vol-I. Scientific Publishers (India), Jodhpur.

Chelladuri, V. 1983. Minnikizhangu- An unique folk medicinal plant from the Adivasis (Tribals) of point Calimere, Tamil Nadu,. *Bull. Medico-ethnobot. Res.* 4:148-153.

Chelvan, P.T. 1998. Traditional Phytotherapy among the migrant Tamilian settlers in South Gujarat. *Biojournal* 10: 9 – 14.

Chetty, K.M. & Rao, K.N. 1989. Ethnobotany of sarakulla & adjacent areas of Chittor district, A.P. *Vegetos* 2: 51-58.

Chetty, K.M., Chetty, M.L. Sudhakar, A. & Ramesh, C. 1998. Ethno-medico botany of some aquatic angiosperm in Chittoor district of A.P., India. *Fitoterapia* 69: 7-12.

Chhetri, D.R. 2005. Ethnomedicinal plants of the Khangchendzonga National Park, *Sikkim, India. Ethnobotany.*17: 96 – 103.

Chopra, R.N. Nayar, S.L. and Chopra, I.C. 1956. Glossary of Indian medicinal plants. CSIR.

Chopra, R.N. Badhwar, R.L. and Ghosh, S. 1965. Poisonous plants of India. Vol. I & II. ICAR, New Delhi.

Dar, G.H., Vir jee, P. Kachroo & G.M. Buth. 1984. Ethnobotany of Kashmir – I. Sind Valley. *J. Econ. Taxon. Bot.*5: 668 – 675.

Das, S.N., K.P. Janardhanan & S.C. Roy. 1983. Some observations on the ethnobotany of the tribes of Totopara and adjoining areas in Jalpaiguri district, West Bengal. *J. Econ. Taxon. Bot.*4: 453 – 474.

Das, S.R. 1995. Some beneficial weeds of cultivation from Nadia district, West Bengal. *Bull. Med. Ethnobot. Res.*16; 74 – 81.

Das, A. K. & G. D. Sharma. 2003. Ethnomedicinal uses of plants by Manipuri and Barman communities of Cachar district, Assam. 421-429. In: Singh *et al.* (ed.). Ethnobotany and Medicinal Plants of India and Nepal. Vol-I. Scientific Publishers (India), Jodhpur.

Das, A. K., B. K. Dutta & G. D. Sharma. 2006. Study of Medicinal plants used by the different tribes/communities of Cachar district Assam. *J. Econ. Taxon. Bot.* 30 (Suppl.): 294-309.

Das, S., T. E. Sheeja & A. B. Mandal. 2006. Ethnomedicinal uses of certain plants from Bay Islands. *IJTK* **5(2)**:207-211.

Dash, S. S., A. Maiti & S. K. Rai. 2003. Traditional uses of plants among the urban population of Gangtok- Sikkim. 317-324. In: Singh *et al.* (ed.). Ethnobotany and Medicinal Plants of India and Nepal. Vol-I. Scientific Publishers (India), Jodhpur.

Datta, M.L. & Nath, S.C. 1998. Ethno-medicobotany of the deories of Assam, India. *Fltoterapia* 69; 147-154.

De, S. & K. M. Parikh. 1985. Medicinal importance of some common Indian vegetables. Part-I. *Anci. Sci. Life.* **4**:232-237.

Deka, D., G. C. Sarma, N. Devi & N. Pathak. 2006. Indigenous herbal medicines used against Malaria in Goalpara and Morigaon district of Assam. *J. Econ. Taxon. Bot.* 30(Suppl.): 177-183.

Deokule, S. S. & D. K. Magdum. 1992. Enumeration of Medicinal plants from Baramati area, district Pune, Maharashtra State. *J. Econ. Taxon. Bot. Addl. Ser.* **10:** 289-299.

Deokule, S. S. 2006. Ethno-medicinal plants of Baramati region of Pune district, Maharashtra. *J. Econ. Taxon. Bot.* 30(Suppl.): 59-69.

Deshmukh, V. R. & S. P. Rothe. 2003. Ethnomedicinal study from Melghat tribal region of Amravati district, Maharashtra. 582-584. In: Singh *et al.* (ed.). Ethnobotany and Medicinal Plants of India and Nepal. Vol-II. Scientific Publishers (India), Jodhpur.

Dhar, U. & Siddique, M.A.A. 1993. Ethnobotanical studies and biological diversity of Suru Valley (Zanskar), *Glimpses Plant Res.* 10: 31-39.

Dhiman, 2003. Sacred plants and their medicinal uses. Daya Publishing House, New Delhi.

Dixit, R. D. & R. Kumar. 2003. Plants used by local people in human welfare. 53-59. In: Singh *et al.* (ed.). Ethnobotany and Medicinal Plants of India and Nepal. Vol-I. Scientific Publishers (India), Jodhpur.

Dobriyal, R.M., Singh, G.S., Rao, K.S. & Saxena, K.G. 1997. Medicinal plant resources in Chhakinal watershed in the North-Western Himalaya, *J Herb Spices Med plant* 5: 15-27.

Duthie, J. F. 1994 (Reprint). Flora of the Upper Gangetic plain. Bishen singh Mahendra Pal Singh, Dehradun.

Dutta Choudhury, M. & S. Shil. 2006. Some traditional antihepatitis plants used by the People of Manipur. *J. Econ. Taxon. Bot.* 30(Suppl.): 289-293.

Dwarakan, P. & A.A. Ansari. 1992. Ethnobotanical notes of Valikadupatti and surroundings of Kollimalais Salem DT., Tamil Nadu. *J. Econ. Taxon. Bot. Addl. Ser.* **10:** 495-499.

Dwivedi, S. N. & Archana Pandey. 1992. Ethnobotanical studies on wild and indigenous species of Vindhyan Plateau. *J. Econ. Taxon. Bot. Addl. Ser.* **10:** 143-150.

Dwivedi, S. N. 2003. Ethnobotanical studies and conservational strategies of wild and natural resources of Rewa district of M. P. 233-244. In: Singh *et al.* (ed.). Ethnobotany and Medicinal Plants of India and Nepal. Vol-I. Scientific Publishers (India), Jodhpur.

Dwivedi, S. N. 2003. Traditionally health care among tribals of Rewa district of Madhya Pradesh with special reference to conservation of endangered and vulnerable species. 315-320. In: Maheshwari, J. K. (ed.). Ethnobotany and Medicinal Plants of Indian Subcontinent. Scientific Publishers (India), Jodhpur.

Dwivedi, S. N., Dwivedi, S. and Patel, P. C. 2005. Herbal remedies used by the rural people in the treatment of joint diseases. *Ethnobotany*. 17: 193-196.

Ganai, K. A. & Irshad A. Nawchoo. 2003. Traditional treatment of toothache by the Gujjar and Bakerwal tribes of Kashmir in India. 105-107. In: Singh *et al.* (ed.). Ethnobotany and Medicinal Plants of India and Nepal. Vol-I. Scientific Publishers (India), Jodhpur.

Ganesan, S. & L. Kesavan. 2003. Ethnomedicinal plants used by the ethnic group valaiyans of Vellimalai hills (Reserved forest) Tamil Nadu. 754-760. In: Singh *et al.* (ed.). Ethnobotany and Medicinal Plants of India and Nepal. Vol-II. Scientific Publishers (India), Jodhpur.

Ganesan S., Venkateshan G & Banumathy N. 2006. Medicinal plants used by ethnic group Thottianaickans of Semmalai hills (reserved forest), Tirucirappalli district, Tamil Nadu. *IJTK*. 5(2): 245-252.

Gantait, S., Kundu, S. K. & Giri, G. S. 2006. Some important medicinal plants of North-East India. *J. Econ. Taxon. Bot.* 30(Suppl.): 283-288.

Gaur, R.D., M.P. Sharma & J.K. Semwal. 1980. Ethnotoxic plants of Garhwal hills in India. *East. Anthrop.*33: 159 – 163.

Ghosh, S. & Sensarma, P. 1997. Ethnomedicine to modern medicine: an observational study in some villages of West Bengal. *Ethnobotany*. 9: 80-84.

Ghosh, R. B. & Debabrata Das. 2003. A preliminary census and systematic survey of antidiabetic plants of Midnapore district, West Bengal. 535-538. In: Maheshwari, J. K. (ed.). Ethnobotany and Medicinal Plants of Indian Subcontinent. Scientific Publishers (India), Jodhpur.

Ghosh, A. 2006. Medicinal plants used for treatment of Diabetes by the tribals of Bankura, Purulia and Medinipur of West Bengal. *J. Econ. Taxon. Bot.* 30(Suppl.):233-238.

Girach, R. D., S. Singh, M. Brahmam & M.K. Misra. 2003. Traditional treatment of skin diseases in Bhadrak district, Orissa. 499-504. In: Maheshwari, J.K. (ed.). Ethnobotany and Medicinal Plants of Indian Sub continent. Scientific Publishers (India), Jodhpur.

Goel, A.K., A.K. Sahoo & V. Mudgal. 1984. A contribution to the ethnobotany of Santal Pargana, Bihar. *Bot. Surv. India.*

Goel, A.K. & V. Mudgal. 1988. A survey of medicinal plants used by the tribals of Santal Pargana (Bihar). *J. Econ. Taxon. Bot.*12: 329 – 335.

Gogoi, P. & C.L. Boissya. 1984. Informations about a few herbal medicines used by the people of Assam (India) against Jaundice, *Himal. Res. Dev.*2: 41 – 44.

Gogoi, R., D. Bokolial & D. Hazarika. 2003. Preliminary observation on the medicinal plants of Chandrapur area of Kamrup District, Assam. 410-415. In: Singh *et al.* (ed.).

Ethnobotany and Medicinal Plants of India and Nepal. Vol-I. Scientific Publishers (India), Jodhpur.

Gogoi, R. & M. K. Das. 2003. Observations on some weeds of Medicinal importance in the Brahmaputra Valley of Assam. 434-441. In: Singh *et al.* (ed.). Ethnobotany and Medicinal Plants of India and Nepal. Vol-I. Scientific Publishers (India), Jodhpur.

Gogte, V.V.M. Vaidya. 2000. Ayurvedic pharmacology and Therapeutic uses of medicinal plants (Dravyaguna Vignyan). Bhartiya Vidya Bhawan, Mumbai.

Gohil, R. N. & M. A. Qadri. 1992. Ethnobotany of Kargil-I medicinal plants used by Balti, Dard & Brokpa races. *J. Econ. Taxon. Bot. Addl. Ser.* **10:** 301-306.

Gopakumar, K. ,Vijayalakshmi, B., Shantha, T. R. & Yoganarasimhan, S.N. 1991. Plants used in Ayurveda from Chikmangalur district, Karnataka – II. *J. Econ. Taxon. Bot.*15: 379 – 389.

Gopal, G.V. & G.L. Shah. 1985. Some folk medicinal plants used for jaundice in Gujarat, India. *J. Res. Edu. Indian Med.*4: 45 – 49.

Gopal, G. V. & K. S. Jagadish Chandra. 2003. Ethnobotany of soligas of Biligirirangana Hills, Karnataka. 640-647. In: Singh *et al.* (ed.). Ethnobotany and Medicinal Plants of India and Nepal. Vol-II. Scientific Publishers (India), Jodhpur.

Gopalakrishnan, K. & V. A. Krishna Prasad. 1992. Some folklore medicine of the tribals (Irulars) of Attappady region (Kerala State). *J. Econ. Taxon. Bot. Addl. Ser.* **10:** 385-380.

Goud, P. S. P., T. Pullaiah & S. R. Murthy. 2003. Native phytotherapy for fever and malaria from kurnool district, Andhra Pradesh. 337-340. In: Maheshwari, J.K. (ed.). Ethnobotany and Medicinal Plants of Indian Subcontinent. Scientific Publishers (India), Jodhpur.

Gupta, O. P., T. N. Srivastava, S. C. Gupta & D. P. Badola. 1980. An ethnobotanical and Phytochemical screening of high altitude plants of Ladakh-I. *Bull. Medico- ethnobot. Res.* 1:301-317.

Gupta, S.P. 1981. Native medicinal uses of plants by the Asurs of Netarhat plateau (Bihar), 218 – 231. In S.K. Jain (Ed.) *Glimpses of Indian Ethnobotany.*

Gupta, Rajendra. 1981. Plants in folk medicine of the Himalayas. 83 – 90. In S.K. Jain (Ed.) *Glimpses of Indian Ethnobotany.*

Gupta, V.C., Hussain, S.J. & Imam, S. 1997. Medico – ethonbotanical survey at Paderu forests of Araku Valley, Andhra Pradesh, *Fitoterapia.*68: 45 – 48.

Gupta, A.K., Mishra, S.K. & Khan, A.A. 1999. Ethnobotanical notes on some herbs from Chhatisgarh region of Madhya Pradesh. *Adv Plant Sci.*12: 163 – 166.

Gupta, A.K. & Tandon, N. 2004 (Ed.). Reviews on Indian Medicinal Plants. *Vol.* I, II, III & IV. *Indian Council of Medical Research. New Delhi.*

Gupta, A.K. 2005 (Ed.). Quality standards of Indian Medicinal Plants. Vol. I, II & III . *Indian Council of Medical Research. New Delhi.*

Gupta, V. 2006. Traditional medicinal plants of the Bangnis of East Kameng district, Arunachal Pradesh. *J. Econ. Taxon. Bot.* 30(Suppl.): 310-319.

Handa, S.S. & Kapoor, V.K. 1999. Pharmacognosy. Vallabh Prakashan, Delhi.

Harja, P.K. & A.K. Baishya. 1981. Ethnobotanical notes on the Miris (Mishings) of Assam Plants, 161 – 169. In S.K. Jain (Ed.): *Glimpses of Inidian Ethnobotany*.

Hemadri, K., Raj, P.V., Rao, S.S. & Sarma, C.R.R. 1980. Folklore clamis from A.P.-I. *J. Sci. Res. Plant Med.* 1: 37-49.

Hemadri, K. & S.S. Rao. 1983. Leucorrhoea and menorrhagia : Tribal medicine. *Anc. Sci. Life.*3: 40 – 41.

Hemadri, K. & S.S. Rao. 1984. Jaundice: Tribal medicine. *Anc. Sci. Life.*3: 209– 212.

Hemadri, K. & Rao, S.S. 1989. Folk-lore claims of Koraput and Phulbani district of Orissa state. *Indian Med.* 1(4): 11-13.

Hemadri, K. & Rao, S.S. B. 1990. Folklore claims of Koraput and Phulbani district of Orissa state. *Indian Med.* 2(1): 4-6.

Hemadri, K. 1992. Tribals of Andhra Pradesh: their knowledge in nutritional and medicinal herbs. *Indian Med.*4(1),1– 6.

Hembrom, P.P. & Goel, A.K. 2005. Horopathy : Ethnomedicine of Mundas. *Ethnobotany* 17: 89 – 95.

Islam, M. A. & R. K. Jha. 2003. Ethnopharmacology of tribals of Ranchi district in Jharkhand. 300-310. In: Singh *et al.* (ed.). Ethnobotany and Medicinal Plants of India and Nepal. Vol-I. Scientific Publishers (India), Jodhpur.

Islam, M. A. & F. Hasin. 2003. Ethnobotany of certain Asteraceous plants of Assam and its neighbouring areas. 442-446. In: Singh *et al.* (ed.). Ethnobotany and Medicinal Plants of India and Nepal. Vol-I. Scientific Publishers (India), Jodhpur.

Jadeja, B. A., N. K. Odedra & S. I. Choudhary. 2006a. Ethnomedicinal plants used by the tribals of Gujarat to cure digestive disorders. *J. Econ. Taxon. Bot.* 30 (Suppl.): 115-123.

Jadeja, B. A., N. K. Odedra, H. N. Mori & D. C. Bhati. 2006b. Medicinal plants of Gujarat used in Gout. *J. Econ. Taxon. Bot.* 30 (Suppl.): 225-232.

Jadeja, B. A., N. K. Odedra & Gajera, M. R. 2006c. Plants used in traditional phytotherapy for hair care in Gujarat. *J. Econ. Taxon. Bot.* 30(Suppl.):258 - 268.

Jadhav, D. 2006. Ethnomedicinal plants used by Bhil tribe of Bibdod, Madhya Pradesh. *IJTK.*5(2): 263 – 267.

Jadhav, D. 2006. Plant sources used for the treatment of types of fevers by Bhil tribe of Ratlam district, Madhya Pradesh. *J. Econ. Taxon. Bot.*30(4): 909– 911.

Jadhav, D. 2006. Ethnomedicinal plants used for curing skin afflictions by Bhil tribe of Ratlam district, Madhya Pradesh. *J. Econ. Taxon. Bot.*30 (Suppl.). 124-127.

Jadhav, D. 2006. Ethno-medicinal survey of Maalgamdi in Ujjain district, Madhya Pradesh, India. *Ethnobotany* 18(1 & 2): 157 – 159.

Jadhav, D. 2007. Ethno-medicinal plants used in the treatment of Diarrhoea and Dysentery by Bhil tribe of Ratlam district (M.P.). *Indian forester*. August:1097- 1100.

Jain, S.K. 1963 a. Studies in Indian ethnobotany – Less known uses of 50 common plants from tribal areas of Madhya Pradesh. *Bull. Bot. Surv. India.*5: 223 – 226.

Jain, S.K. 1963 b. Studies in Indian ethnobotany – II. Plants used in medicine by the tribals of Madhya Pradesh. *Bull. Reg. Res. Lab. Jammu.*1: 126 128.

Jain, S.K. 1965 a. Medicinal plantlore of the tribals of Bastar. *Econ. Bot.*19: 236 -250.

Jain, S.K. 1965 b. On the prospects of some new or less known medicinal plant resources. *Indian Med. J.* 59: 270 – 272.

Jain, S.K. & J.N. De. 1966. Observations on ethnobotany of Purulia district, West Bengal. *Bull. Bot. Suv. India.* 8: 237 – 251.

Jain, S.K. & C.R. Tarafder. 1970. Medicinal plant lore of the Santals. A revival of P O Bodding's work. *Econ. Bot.* 24: 241 – 278.

Jain, S.K., D.K. Banerjee & D.C. Pal. 1973. Medicinal plants among certain Adibasis in India. *Bull. Bot. Suv. India.* 15: 85 – 91.

Jain, S.K. & N. Dam. 1979. Some ethnobotanical notes from north – eastern India. *Econ. Bot.* 33: 52 – 56.

Jain, S.K. & S.K. Borthakur. 1980. Ethnobotany of the Mikirs of India. *Econ. Bot.* 34: 264 – 272.

Jain, S.K. 1981 (Ed.). Glimpses of Indian ethnobotany. Oxford and IBH publishing Co., New Delhi

Jain, S.K. 1984. Ethnobotany of Morni and Kalesar (Ambala – Haryana). *J. Econ. Taxon. Bot.* 5: 809 – 813.

Jain, S.K. 1991. Dictionary of Indian folk Medicine & ethnobotany. Deep publication, New Delhi.

Jain, P. & Sahu, T.R. 1993. An ethnobotanical study of Noradehi sanctuary park of M.P., India : Native plant remedies for Scorpion sting & Snake bite. *J. Econ. Taxon. Bot.* 17: 315-328.

Jain, S.P. & Puri, H.S. 1994. A ethno-medico botanical survey of Parbati valley in H.P. (India), *J Econ. Tax. Bot.* 18: 321-327.

Jain, S. K. 1997. Contribution to Indian Ethnobotany. Scientific Publishers (India), Jodhpur.

Jain, S. P., S. C. Singh, J. Singh & Sushil Kumar. 2003. Ethno-medico-botanical survey of Raipur district, Chhattisgarh State. 266-271. In: Singh *et al.* (ed.). Ethnobotany and Medicinal Plants of India and Nepal. Vol-I. Scientific Publishers (India), Jodhpur.

Jain, S.K. 2004 (Ed.). A manual of ethnobotany. Scientific publisher, Jodhpur.

Jamir, N.S. 1990. Some interesting medicinal plants used by Nagas. *J. Res. Edue Indian Med.* 9: 81-87.

Janardhanan, K.P. 1963. An enumeration of the medicinal plants of Khed Taluka, Maharashtra State. *Bull. Bot. Surv. India.*5: 363 – 374.

Janmeda, B. S., Lokendra Singh & Preeti Vats. 2006. Folk claims on Monocotyledons plants in East- Sikkim. *J. NTFPs* **13(2):15**5-160.

Jha, P.K. Choudhary, R.S. & Choudhary, S.K. 1997. Studies of Medicinal plants of Palamau (Bihar)- (Part – II). *Bijournal.* 9: 21 – 38.

Jhon, D. 1984. One hundred useful raw drugs of the Kani tribes of Trivandrum forest division, Kerala, *Int J Crude Drugs Res.* 22, 17-39.

Joshi, M.C., M.B. Patel & P.J. Mehta. 1980. Some folk medicines of Dangs, Gujarat State. *Bull. Medico – ethnobot. Res.* 1: 8 – 24.

Joshi, P. 1982. An ethnobotanical study of Bhils – A preliminary survey. *J. Econ. Taxon. Bot.*3: 257 – 266.

Jubilee, P. & Subhan, C. N. 2006. Biological activities of ethnomedicinal claims of some plant species of Assam. *IJTK.*5(2):229-236.

Kadavul, K. & P. Jisha. 2006. Medicinal plants and their conservation in Mahe region, U. T. of Pondicherry, Malabar Coast, South India.. *J. Econ. Taxon. Bot.* 30(Suppl.): 203-207.

Kala, C. P. & Nima Manjrekar. 2003. Ethnomedicobotany of Indian trans-Himalaya: A case study of Spiti. 177- 183. In: Maheshwari, J. K. (ed.). Ethnobotany and Medicinal Plants of Indian Subcontinent. Scientific Publishers (India), Jodhpur.

Kataria, R. L. & D.P. Gauchan. 2006. Ethnomedicinal plants used in Dysentery and diarrhoea in Hanumangarh district, Rajasthan. *J. Econ. Taxon. Bot.* 30(Suppl.): 128-132.

Kamble, S.Y. & S.G. Pradhan. 1980. Ethnobotany of Korkus in Maharashtra. *Bull. Bot. Surv. India.* 22: 201 – 202.

Kapoor, S.L. & Kapoor, L.D. 1980. Medicinal plant wealth of the Karimnagr district of A.P. *Bull. Med. Ethnobot. Res.* 1: 120-144.

Kapur, S. K. & S. Nanda. 1992. Traditionally important medicinal plants of Bhaderwah Hills- Jammu Province - I. *J. Econ. Taxon. Bot. Addl. Ser.* **10:** 307-318.

Kapur, S. K., S. Nanda & Y. K. Sarin. 1992. Ethnobotanical uses of RRL Herbarium-I. *J. Econ. Taxon. Bot. Addl. Ser.* **10:** 461-493.

Kapur, S.K. 1993. Ethno – Medico plants of Kangra Valley (Himachal Pradesh). *J. Econ. Taxon. Bot.*17: 395 – 408.

Karnick, C.R., K.C. Tiwari, R. Majumdar & S. Bhattacharjee. 1981. Newer ethnobotanical and folklore studies of some medicinal plants of Gauhati and surrounding areas. *Nagarjun.*24: 240 – 245.

Katewa, S.S. & Arora, A. 1997. Some plants in folk medicines of Udaipur district. (Rajasthan). *Ethnobotany.* 9: 48 – 51.

Katewa, S.S. & Sharma, R. 1998. Ethnomedicinal observations from certain Watershed areas of Rajasthan. *Ethnobotany.* 10: 46 – 49.

Khan, A. A., S. Khan & M. P. Singh. 2003. Weeds of Soyabean [*Glycine max* (L.) Merr.] and their medicinal importance in Rewa division (M.P.). 245-248. In: Singh *et al.* (ed.). Ethnobotany and Medicinal Plants of India and Nepal. Vol-I. Scientific Publishers (India), Jodhpur.

Khan, A.A., A.K. Verma & M.P. Singh. 2003. Certain ethnobotanical informations on food and medicinal plants from Rewa division of Madhya Pradesh with special reference to rare and endangered species. 249-254. In: Singh *et al.* (ed.). Ethnobotany and Medicinal Plants of India and Nepal. Vol-I. Scientific Publishers (India), Jodhpur.

Kharkongor, P. & J. Joseph. 1981. Folklore medico – botany of rural Khasi and Jaintia tribes in Meghalaya. 124 – 136. In S.K. Jain (Ed.) *Glimpses of Indian Ethnobotany.*

Khory, J.K. 1981. Materia medica of India & their Therapeutics, Neeraj Publishing house, New Delhi.

Kothari, M. J. & K. M. Rao. 2003. Ethnobotanical studies of Thane district, Maharashtra. 265-272. In: Maheshwari, J. K. (ed.). Ethnobotany and Medicinal Plants of Indian Subcontinent. Scientific Publishers (India), Jodhpur.

Kothari, M. J. & A. N. Londhe. 2003. Ethnobotany in Human health care of Chikhaldara, Amaravati district in Maharashtra State. 273-281. In: Maheshwari, J.K. (ed.). Ethnobotany and Medicinal Plants of Indian Subcontinent. Scientific Publishers (India), Jodhpur.

Kirtikar, K.R. & Basu, B.D. 2004. Indian medicinal plants, Vol. I, II, III &IV (Reprint). Bishen Singh Mahendra Pal Singh, Dehradun.

Kiruba, S., S. Jeeva, N. Venugopal, S.S.M. Das, G.S. Regini, R.C. Laloo and B.P. Mishra. 2006. Ethnomedicinal herbs of Koonthakulam water Sanctuary, Nellai, Tamilnadu, India. *J. NTFPs* **13(1)**:25-27.

Krishna, B. & S. Singh. 1987. Ethnobotanical observations in Sikkim. *J. Econ. Taxon. Bot.* 9: 1 – 7.

Kshirsagar, R.D. & N.P. Singh. 2003. Less-known herbal remedies from Soligas of Mysore, Southern India. 656- 661. In: Singh *et al.* (ed.). Ethnobotany and Medicinal Plants of India and Nepal. Vol-II. Scientific Publishers (India), Jodhpur.

Kumar, Y., K. Haridasan & R.R. Rao. 1980. Ethnobotanical notes on certain medicinal plants among some Garo people around Balphakram Sanctury in Meghalaya. *Bull. Bot. Surv. India.*22: 161 – 165.

Kumar, Y., S. Fancy & R.R. Rao. 1987. Further contribution to the ethnobotany of Meghalaya: Plants used by War Jaintia of Jaintia Hill district. *J. Econ. Taxon. Bot.*11: 65 – 70.

Kumar, M. 1992. Ethnomedicinal studies of plants of chotanagpur plateau, Bihar, *Int. J. Mendel.9: 137 – 139.*

Kumar, V. & Jain, S.K. 1998. A contribution to ethnobotany of Surguja district in M.P. *Ethnobotany.*10: 89 – 96.

Kumar, K. & Goel, A. K. 1998. Little known ethnomedicinal plants of Santhal & Paharia tribes in Santhal Paragna, Bihar, India. *Ethnobotany.*10: 66-69.

Kumar, S., F. Parveen & R. S. Mertia. 2003. Taxonomic analysis of medicinal plants of the Indian Thar: A Prioritisation strategy. 160-169. In: Singh *et al.* (ed.). Ethnobotany and Medicinal Plants of India and Nepal. Vol-I. Scientific Publishers (India), Jodhpur.

Kumar, Anand & Jagdish Lal. 2003. Medicinal plant resources of the Kanha tiger reserve, M.P. 255-265. In: Singh *et al.* (ed.). Ethnobotany and Medicinal Plants of India and Nepal. Vol-I. Scientific Publishers (India), Jodhpur.

Kumar, K.J.L., K.S. Kalesh & A. Mohandas. 2003. An account of the Economic potential and commercial exploitation of four lesser known *Garcinia* species from India. 692-

694. In: Singh *et al.* (ed.). Ethnobotany and Medicinal Plants of India and Nepal. Vol-II. Scientific Publishers (India), Jodhpur.

Kumar, A., D.D. Tewari & Y.N. Pande. 2003. Ethno-phytotherapeutics among Tharus of Beerpur-Semra forest range of Balrampur district (U.P.). 839-844. In: Singh *et al.* (ed.). Ethnobotany and Medicinal Plants of India and Nepal. Vol-II. Scientific Publishers (India), Jodhpur.

Kumar, S. & Chauhan, A.K.K. 2005. Medicinal plants used by local inhabitants in Bharatpur district, Rajasthan. *Ethnobo.*17: 179 – 183.

Kumar, A., A. Singh, Y. Sharma & J. C. Rana. 2006. Ethno-medico-botany of some important plants in Mandhala watershed of Himachal Pradesh. *J. Econ. Taxon. Bot.* 30(Suppl.): 145-150.

Kumar, A. & D. K. Yadav. 2006. Important Ethno-medicinal plants of family Poaceae in Gaya District, Bihar. *J. Econ. Taxon. Bot.* 30(Suppl.): 171-176.

Kumar, T. D. C. & Pullaiah. 2003. Ethnomedicinal uses of some plants of study of Mahabubnagar district, Andhra Pradesh, India. 341-345. In: Maheshwari, J. K. (ed.). Ethnobotany and Medicinal Plants of Indian Subcontinent. Scientific Publishers (India), Jodhpur.

Lakshamanan, K.K. & A.S. Sankaranarayanan. 1988. Some folklore medicines in the remote hemlets, Dhoomanoor & Champukarai of Anai Katty hills, Coimbatore, Tamil Nadu. *Indian J. Forestry.*11: 217 – 219.

Lal, S.D. & B.K. Yadav. 1983. Folk Medicines of Kurukshetra District (Haryana), India. *Econ. Bot.* 37: 299 – 305.

Lal, S.D. & K. Lata. 1980. Plants used by the Bhat Community for regulating fertility. *Econ. Bot.* 34: 273 – 275.

Lindley, John. 1981. Flora medica, Ajay Book Service, New Delhi.

Maheshwari, J. K., K. K. Singh & S. Saha. 1980. Ethnomedicinal uses of plants by Tharus in Kheri district, U. P. *Bull. Medico-ethnobot. Res.* 1:318-337.

Maheshwari, J.K. & K.K. Singh & S. Saha. 1981. The Ethonbotany of the Tharus of Kheri district. Uttar Pradesh. *Nation. Bot. Res. Inst.* Lucknow.

Maheshwari, J.K. & J.P. Singh. 1984. Contribution to the ethnobotany of Bhoxa tribe of Bijnor & Pauri Garhwal district, U.P. *J. Econ. Taxon. Bot.* 5: 251 – 259.

Maheshwari, J.K. B.S. Kalakoti & Brijlal. 1986. Ethnomedicine of Bhil Tribe of Jhabua district, Madhya Pradesh. *Anc. Sci. Life.* 5: 255 – 261.

Maheshwari, J.K. K.K. Singh & S. Saha. 1986. Ethnobotany of tribals of Mirzapur district. U.P. *Nation. Bot. Res. Inst.*, Lucknow.

Maheshwari, J.K. & J.P. Singh. 1987. Traditional phytotherapy amongst the Kol tribe of Banda district, Uttar Pradesh. *J. Econ. Taxon. Bot.* 9: 165 – 171.

Maikhuri, R. K. & P. S. Ramakrishnan. 1992. Ethnobiology of some tribal societies of Arunachal Pradesh in North-East India. *J. Econ. Taxon. Bot. Addl. Ser.* **10:** 61-78.

Maikhuri, R. K., Nautiyal, S., Rao, K. S. & K. G., Saxena. 1998. Medicinal plants cultivation and Biosphere reserve management: A case study from Nanda devi Biosphere reserve, Himalaya. *Curr. Sci.* **74:**157-163.

Malhotra, S.K. & S. Moorthy. 1973. Some useful and medicinal plants of Chandrapur district (Maharasthra State). *Bull. Bot. Surv. India.*15: 13 – 21.

Malhotra, C.L. & Balodi, B. 1984. Medicinal plants of Alpine region of Gori Valley, Kumaon, *J. Sci. Res. Plant Med* 5(1&2): 1-8.

Manandhav, N. P. 2003. Ethnobotanical notes on the family Leguminoseae in Nepal. 333-344. In: Singh *et al.* (ed.). Ethnobotany and Medicinal Plants of India and Nepal. Vol-I. Scientific Publishers (India), Jodhpur.

Megoneitso & R.R. Rao. 1983. Ethnobotanical studies in Nagaland – 4. Sixty two medicinal plants used by the Angami Nagas. *J. Econ. Taxon. Bot.*4: 167 – 172.

Meena, S. L., K. C. Sharma & R. Gopalan. 2003. Ethnomedicinal plants of Karauli district, Rajasthan. 177-180. In: Singh *et al.* (ed.). Ethnobotany and Medicinal Plants of India and Nepal. Vol-I. Scientific Publishers (India), Jodhpur.

Mishra, R. & R.D. Dxit. 1973. Vajradanti – An ethnobotanical study. I. *J. Res. Indian Med.*8: 89 – 91.

Mishra, R. & R. D. Dixit. 1976. Studies on Ethnobotany of some less known plants of Ajmer forest division, Rajasthan. *Nagarjun.* 19: 20-22.

Mishra, R. & K.V. Billore. 1983. Some ethnobotanical lores from Banswara district. Rajasthan. *Nagarjun.*26: 229 – 231.

Mishra, D.P. & T.R. Sahu. 1984. Euphorbiaceous plants used in medicine by the tribals of Madhya Pradesh, India. *J. Econ. Taxon. Bot.*5: 791 – 794.

Mishra, S.K. & Verma, R.M.P. 1995. Medicinal plants of Champaran (Bihar) used in various diseases. *Flora fauna.*1: 75 – 76.

Mishra, B. & Vaisya, R. 2004. Bhavaprakasa, Vol. I & II. Chaukhambha Sanskrit Sansthan, Varanasi.

Mitaliya, K. D., D. C. Bhatt, J. N. Patel & S.K. Patel. 2003. Observations on plants used for joint disorders by rural and tribal people of Gujrat. 212-216. In: Singh *et al.* (ed.). Ethnobotany and Medicinal Plants of India and Nepal. Vol-I. Scientific Publishers (India), Jodhpur.

Mitaliya, K. D. & D. C. Bhatt. 2003. Rare and endangered medicinal plants of Gujrat. 845-850. In: Singh *et al.* (ed.). Ethnobotany and Medicinal Plants of India and Nepal. Vol-II. Scientific Publishers (India), Jodhpur.

Mohanty, R.B. Padhy, S.N. & Dash, S.K. 1996. Traditional phytotherapy for diarrhoeal diseases in Ganjan and Phulbani district of South Orissa, India. *Ethnobotany.*8: 60 – 65.

Mokat, D. N. & S. S. Deokule. 2006. Ethno-medico-botanical survey of Ratnagiri district of Maharashtra State. *J. Econ. Taxon. Bot.* 30(Suppl.): 140-144.

Molla, H.A. & B. Roy. 1985. Traditional uses of some medicinal plants by the Rabha tribals in Jalpaiguri district, West Bengal. *J. Econ. Taxon. Bot.*7: 578 – 580.

Molla, H. A. & D. C. Pal. 1992. Observation on *Vetiveria zizanioides* (L.) Nash. (Khas-Khas). *J. Econ. Taxon. Bot. Addl. Ser.* **10:** 101-103.

Mondal, P. & P. K. Mukherjee. 1992. Notes on Ethnobotany of Keonjhar district, Orissa. *J. Econ. Taxon. Bot. Addl. Ser.* **10:** 7-18.

Mudgal, V. & D.C. Pal. 1980. Medicinal plants used by tribals of Mayurbhanj (Orissa). *Bull. Bot. Surv. India.* 22: 59 – 62.

Mudgal, V. Khanna, K.K. & Hajra, P.K. 1997. Flora of Madhya Pradesh. Vol. II. BSI. Calcutta.

Mukerji, B. 1953. The Indian pharmaceutical Codex, New Delhi.

Mukherjee, A.S. Mukherjee & M. Mukherjee. 1986. Less known medicinal plants of Rubiaceae in Darjeeling Hills, West Bengal (India). *Indian J. Appl. & Pure. Bot.*1: 86 – 91.

Mukhopadhyay, C. R. 1992. Useful plants of Birbhum district, West Bengal. *J. Econ. Taxon. Bot. Addl. Ser.* **10:** 83-95.

Murthy, K.S., P.C. Sharam & Prem Kishore. 1986. Tribal remedies for sanke bite from Orissa. *Anc. Sci. Life.*6: 122 – 123.

Muthukumarasamy, S., V. R. Mohan, S. Kumaresan & V. Chelladurai. 2003. Herbal medicinal plants used by Palliyars to obtain relief from gastro-intestinal complaints. 711-714. In: Singh *et al.* (ed.). Ethnobotany and Medicinal Plants of India and Nepal. Vol-II. Scientific Publishers (India), Jodhpur.

Nadanakunjidam, S. 2003. Ethnomedicinal observation from Attapadi Hills of Western Ghats. 732-740. In: Singh *et al.* (ed.). Ethnobotany and Medicinal Plants of India and Nepal. Vol-II. Scientific Publishers (India), Jodhpur.

Nadkarni, K. M. 2002 (Reprint). Indian Materia medica, Vol.-I. Bombay popular prakashan.

Nagar, M. 1985. The use of wild plant foods by aboriginal communities in Central India, 337 – 342. In Mishra, V.M. & Belwood, P. (Ed.) *Recent Adv. In Indo-Pacific Prehist.* Oxford & IBH, New Delhi.

Nagarjun, N. & Rao, K.N. 1989. Folk - medicine for diabetes from Rayalaseema of Andhra Pradesh, *Ancient. Sci. Life.*9: 31 – 35.

Nagendra Prasad, P. & Z. Abraham. 1984. Ethnobotany of Nayadis of North Kerala. *J. Econ. Taxon. Bot.*5: 41 – 48.

Namjoshi, A.N. 1960. Studies in Pharmacognosy of Ayurvedic Drugs. Board of Research on Ayurveda, Bombay.

Narain, S. & Singh, U. 2006a. Contribution to the Ethnobotanical plants of Mirzapur district (U. P.). *J. Econ. Taxon. Bot.* 30(Suppl.): 14 – 17.

Narain, S. & Singh, U. 2006b. Contribution to the Ethnobotanical plants of Sonbhadra district (U. P.). *J. Econ. Taxon. Bot.* 30(Suppl.): 18 – 20.

Nath, Ashis & G. G. Maiti. 2003. Ethnobotany of Barak valley (Southern Assam) with special reference to folk medicine. 964-971. In: Singh *et al.* (ed.). Ethnobotany and Medicinal Plants of India and Nepal. Vol-II. Scientific Publishers (India), Jodhpur.

Nath, P. C., Sarmah, R., Arunachalam, A. & Barbhuiya, A. 2006. NTFPs: Potential and their conservation in Dihing-patkai wildlife santuary of Assam. *Jou. Of NTFPs.* 13: 17-23.

Nautiyal, S. 1981. Some medicinal plants of Garhwal hills – A traditional use. *J. Sci. Res. Pl. Med.*2: 12 – 18.

Nautiyal, S., R. K. Maikhuri, K. S. Rao & K. G. Saxena. 2003. Ethnobotany of the tolchha bhotiya tribe of the buffer zone village in Nanda biosphere reserve, India. 119-142. In: Singh *et al.* (ed.). Ethnobotany and Medicinal Plants of India and Nepal. Vol-I. Scientific Publishers (India), Jodhpur.

Nayak, P. K., R. K. Nayak & B. P. Choudhury. 2003. A check list of medicinal plants of Kalahandi district in Orissa. 519-532. In: Singh *et al.* (ed.). Ethnobotany and Medicinal Plants of India and Nepal. Vol-II. Scientific Publishers (India), Jodhpur.

Nayak, R. K., P. K. Nayak & B. P. Choudhury. 2003. Some medicinal weeds of Mahanadi Delta. 533-538. In: Singh *et al.* (ed.). Ethnobotany and Medicinal Plants of India and Nepal. Vol-II. Scientific Publishers (India), Jodhpur.

Negi, K.S., J.K. Tiwari & R.D. Gaur. 1985. Economic importance of some common trees in Garhwal, Himalaya: An ethnobotanical study. *Indian J. Forestry.* 8: 276 – 289.

Oommachan, M. 1977. The flora of Bhopal. J. K. Jain Brothers, Bhopal.

Pal, D.C. & J.N. Srivastava. 1976. Preliminary notes of ethnobotany of Singhbhum district, Bihar, *Bull. Bot. Surv. India.*18. 247 – 250.

Pal, D.C. 1980. Observations on folklore about plants used in veterinary medicine in Bengal, Orissa & Bihar. *Bull. Bot. Surv. India.* 22: 96 – 99.

Pal, G.D. 1984. Observations on ethnobotany of tribals of Subansiri. Arunachal Pradesh. *Bull. Bot. Surv. India.* 26: 26 - 37.

Pal, D.C. & S.K. Jain. 1989. Notes on Lodha medicine in Midnapur district. W.B. India. *Econ. Bot.* 43: 464 – 470.

Pal, D. C., A. Guha & R. Sen. 2003. Medicinal plants and plant products used in children diseases among aboriginals in India. 121-130. In: Maheshwari, J.K. (ed.). Ethnobotany and Medicinal Plants of Indian Subcontinent. Scientific Publishers (India), Jodhpur.

Panda, P. C. & P. Das. 2003. Medicinal plant-lore of the tribals of Baliguda sub-division, Phulbani district, Orissa. 515-521. In: Maheshwari, J.K. (ed.). Ethnobotany and Medicinal Plants of Indian Subcontinent. Scientific Publishers (India), Jodhpur.

Pandey, G.S. 1969. Bhavaprakash Nighantu (With Commentary by Dr. K.C. Chunekar). Chowkhamba Sanskrit Sansthan, Varanasi.

Pandey, A.K. & S.D. Rout. 2006. Ethnobotanical uses of plants of Similipal Biosphere Reserve (Orissa). *Ethnobotany* 18: 102-106.

Pandit, B.R. Kotiwar, O.S., Oza, R.A. & Kumar, R.M. 1996. Ethno - medicinal plant lore from Gir forest, Gujarat. *Adv Plant Sci.* 9: 81 – 84.

Panigrahi, A.K. & Alka, S. 2000. Glossary of useful and economically important plants. New Central Book Agency. Calcutta.

Paranjpe, P. Indian medicinal plants: forgotten healers. 2005. Chaukhamba Sanskrit Pratisthan, Delhi.

Pawar, S. & Patil, D. A. 2006. Ethnobotanical aspects of naturalised exotic plants of Jalgaon district, Maharashtra. *Journal of NTFPs.* 13:55-59.

Pearsall, J. & Trumble, B. 1996. The oxford english Reference Dictionary. Oxford University Press, Oxford.

Pereira, N. 2006. Ethnobotanical relevance in tribal life: A study on Warli tribe of Thane district, Maharashtra. *J. Econ. Taxon. Bot.* 30(Suppl.): 31-39.

Peter Paul Hembrom and Goel, A.K. 2005. Horopathy: Ethnomedicine of Mundas. *Ethnobotany.* 17:89-95.

Prakash, A. & K. K. Singh. 2006. Useful medicinal plants for general tonic, sexual vigour, strength, vitality and venereal diseases among the tribals of Uttar Pradesh and Uttaranchal.. *J. Econ. Taxon. Bot.* 30(Suppl.): 326-331.

Prasad, V. K., T. Rajagopal & K. V. S. Badarinath. 2003. Notes on economic importance of wild plants of Rampa agency- East Godovari district, Andhra Pradesh. 603-612. In: Singh *et al.* (ed.). Ethnobotany and Medicinal Plants of India and Nepal. Vol-II. Scientific Publishers (India), Jodhpur.

Printya, J.V. 1963. The Sushruta Samhita. Vol. I & II, Motilal Banarsidas. Delhi.

Puri, H.S. 1983. Medicinal plants of Tejpur (Assam). *Bull. Medico-ethnobot. Res.*4: 1-13.

Purkayastha, J. & Subhan, C.N. 2005. Biological activitis of ethnomedicinal claims of some plant species of Assam. *IJTK.* 5(2), 229 – 236.

Purohit, V.P., R.A. Silas and R.D. Gaur. 1985. Ethnobotanical studies of some medicinal plants used in skin diseases from Raath (Pauri) Garhwal, Himalaya. *J. Sci. Res. Pl. Med.*6: 39 – 47.

Radhakrisnan, K., Pandurangan, A.G., Pusphangadan, P. & Sasidharan, A. 1996. Less known ethnomedicinal plants of Kerala state and their conservation. *Ethnobotany* 8: 82-84.

Rai, M.K. 1985. Plants used as medicine by tribals of Chhindwara District (M.P.). *J. Econ. Taxon. Bot.*7: 385 – 387.

Rai, Bharat, D. B. Khawas & Rabindra Chettri. 1992. Certain plants in the folklore and folklife of Darjeeling and Sikkim Hills. I: Used for the treatment of human ailments. *J. Econ. Taxon. Bot. Addl. Ser.* **10:** 193-198.

Rajan, S. 1992. Medicinal Plants of Ootacamund, Nilgiris, Tamil Nadu. *J. Econ. Taxon. Bot. Addl. Ser.* **10:** 429-460.

Rajan, S., D. Suresh Baburaj, M. Sethuraman & S. Parimala. 2003. Folk-Medicine among Nilgiri Irulas: An emerging trend in primary health care practice. 791-795. In: Singh *et al.* (ed.). Ethnobotany and Medicinal Plants of India and Nepal. Vol-II. Scientific Publishers (India), Jodhpur.

Rajan, S., H. C. Gupta & S. Kumar. 2003. Exotic medicinal plants of Lucknow. 205-222. In: Maheshwari, J.K. (ed.). Ethnobotany and Medicinal Plants of Indian Sub-continent. Scientific Publishers (India), Jodhpur.

Rajendra, S. M. & Mehrotra, B.N. 1996. Unrecorded medicinal uses of plants among parambikulam tribals, Kerala, India. In S. K. Jain (Edited) Ethnobiology in human welfare, Deep publications, Delhi. Pp. 181-183.

Rajendran, S. M. 2003. West Sikkim: A treasure house of medicinal plants and their utility by local tribals. 311-316. In: Singh *et al.* (ed.). Ethnobotany and Medicinal Plants of India and Nepal. Vol-I. Scientific Publishers (India), Jodhpur.

Rajendran, S. M. & B. S. Aswal. 2003. Some flowering plants used as cosmetics among tribals of Nilgiris, Tamilnadu. 425-430. In: Maheshwari, J. K. (ed.). Ethnobotany and Medicinal Plants of Indian Subcontinent. Scientific Publishers (India), Jodhpur.

Rajendran, K., R. Balakrishnan & S. Chandrasekaran. 2003. Common medicinal plants and their utilization by villagers in east coast. 727-731. In: Singh *et al.* (ed.). Ethnobotany and Medicinal Plants of India and Nepal. Vol-II. Scientific Publishers (India), Jodhpur.

Rajendran, K. & S. K. Rengamani. 2006. Medicinal plants and their utilization by villagers in Southern districts of Tamilnadu. *J. Econ. Taxon. Bot.* 30(Suppl.): 208-216.

Rajwar, G.S. 1983. Low altitude medicinal plants of South Garhwal. *Bull. Medico-ethnobot. Res.*4: 14 – 28.

Ramachandran, V.S. & N.C. Nair. 1981 a. Ethnobotanical studies in cannanore district, Kerala state (India). *J. Econ. Taxon. Bot.* 2: 65 - 72.

Ramachandran, V.S. & N.C. Nair. 1981 b. Ethnobotanical observations on 'lrulars' of Tamil Nadu (India). *J. Econ. Taxon. Bot.* 2: 183 – 190.

Rana, T.S., Datt. & Rao, R.R. 1996. Strategies for sustainable utilization of plant resources by the tribals of the Tons - Valley, Western Himalayas. *Ethnobotany.*8: 96-104.

Rao, R.R. & B. Neogi, 1980. Observations on the ethnobotany of the Khasi & Garo tribes in Meghalaya. *J. Econ. Taxon. Bot.* 1: 157 – 162.

Rao, R.R. 1981 a. Ethnobotanical stuidies on the flora of Meghalaya – Some interesting reports of herbal medicines. 137-148. In S.K. Jain (Ed.) *Glimpses of Indian Ethnobotany.*

Rao, R.R. 1981b. Ethnobotany of Meghalaya – Medicinal plants used by Khasi and Garo tribes. *Econ. Bot.*35: 4 – 9.

Rao, R.R. & N.S. Jamir. 1982a. Ethnobotanical studies in Nagaland – II: 54 Medicinal plants used by Nagas. *J. Econ. Taxon. Bot.*3: 11 – 17.

Rao, R.R. & N.S. Jamir, 1982b. Ethnobotanical studies in Nagaland – I: Medicinal plants. *J. Econ. Bot.*36: 176 – 181.

Rao, K.P. & Sreeramulu, S.H. 1985. Ethnobotany of selected medicinal plants of Srikakulam district, A.P. *Ancient Sci. life.* 4: 238-244.

Rao, R.R. & Haridasan, K. 1991. An ethnobotanical survey of medicinal and other useful plants form North-East India. *J. Eco. Tax. Bot.* 15: 423-436.

Rao, V.L.N., B. R. Busi, B. Dharma Rao, Ch Seshagiri Rao, K Bharathi & M Venkaiah. 2006. Ethnomedicinal practices among Khonds of Visakhapatanam district, A.P. *IJTK* **5(2):**217-219.

Rastogi & Mehrotra 1997. Compendium of Indian medicine Plant, Vol. I, II, III, IV & V. CDRI, Lucknow and PID Delhi.

Ravikumar, K. & R. Vijaya Sankar. 2003. Ethnobotany of Malayali tribes in Melpattu village, Javvadhu Hills of Eastern ghats, Tiruvannamalai district, Tamil Nadu. In: Singh *et al.* (ed.). Ethnobotany and Medicinal Plants of India and Nepal. Vol-II. Scientific Publishers (India), Jodhpur.

Rawat, G. s. & Y. P. S. Pangtey. 1987. A contribution to the ethnobotany of Alpine region of Kumaon. *J. Econ. Taxon. Bot.* 11:139-148.

Reddy, M.B., Reddy, K.R. & Reddy, M.N. 1991. Ethnobotany of Cuddapah district. A.P., India. *Int. J. Pharmacog.* 29; 273-280.

Rosakutty, P. J., A. Stella Roslin & S. Ignacimutha. 2003. Sone traditional folklore medicinal plants of Kanyakumari district, Tamil Nadu. 369-375. In: Maheshwari, J. K. (ed.). Ethnobotany and Medicinal Plants of Indian Subcontinent. Scientific Publishers (India), Jodhpur.

Roy, G.P. & K.K. Chuturvedi. 1986a. Plants used as tonic by the tribals of Abujh Marh Reserve area, Bastar district in Madhya Pradesh. *Vanyajati.*34: 27 – 32.

Roy, G.P. & K.K. Chaturvedi. 1986b. Ethnomedicinal trees of Abujh-Marh area, Madhya Pradesh. *Folklore.*27: 95 – 100.

Roy, G.P. & K.K. Chaturvedi. 1987. Less known medicinal uses of rare and endangered plants of Abujh Marh reserve area, Bastar (Madhya Pradesh). *J. Econ. Taxon. Bot.* 9: 325 – 328.

Sadhale, A., Mujumdar, A.M. & Pendse, G.S. 1991. Ethnobotanical studies of sacred grove at Ajiwali, Pune district, *J. Econ. Taxon. Bot.* 15: 167-172.

Saikia, N. & S. C. Nath. 2003. Ethnomedicinal observations of some species of the genus *Curcuma* L. growing in Assam. 430-433. In: Singh *et al.* (ed.). Ethnobotany and Medicinal Plants of India and Nepal. Vol-I. Scientific Publishers (India), Jodhpur.

Saikh, N. & Subhan C. Nath. 2003. Indigenous practice of treating Dental disorders in Assam. 496-499. In: Singh *et al.* (ed.). Ethnobotany and Medicinal Plants of India and Nepal. Vol-I. Scientific Publishers (India), Jodhpur.

Saikia, N. & S. C. Nath. 2003. Traditional use of some plants in religio-cultural purposes by Koch community in Assam, India. 997-1001. In: Singh *et al.* (ed.). Ethnobotany and Medicinal Plants of India and Nepal. Vol-II. Scientific Publishers (India), Jodhpur.

Sahoo, A. K. & D. D. Bahali. 2003. Plants used in less known ethnomedicines and medico-religious beliefs in Phulbani district of Orissa. 500-504. In: Singh *et al.* (ed.). Ethnobotany and Medicinal Plants of India and Nepal. Vol-I. Scientific Publishers (India), Jodhpur.

Samvatsar, S. & V. B. Diwanji. 2003. Plants used for Rheumatism by the tribals of Western M. P. 305-314. In: Maheshwari, J. K. (ed.). Ethnobotany and Medicinal Plants of Indian Subcontinent. Scientific Publishers (India), Jodhpur.

Saren, A. M., R. Sen & D.C. Pal. 2003. A contribution to the ethnobotany of Bankura district, West Bengal. 545-555. In: Maheshwari, J. K. (ed.). Ethnobotany and Medicinal Plants of Indian Subcontinent. Scientific Publishers (India), Jodhpur.

Sarkar, N., S. Rudra & S. K. Basu. 2003. Ethnobotany of Bangriposi, Mayurbhanj, Orissa. Traditional medicinal uses of plants of Himachal Hills. 509-514. In: Maheshwari, J. K. (ed.). Ethnobotany and Medicinal Plants of Indian Subcontinent. Scientific Publishers (India), Jodhpur.

Sarma, G. C. & R. S. Baishya. 1992. Studies on the medicinal uses of plants by the North-East tribes-IV. *J. Econ. Taxon. Bot. Addl. Ser.* **10:** 177-184.

Sarma, S. K., B. Bevi & D. K. Bhattacharjya. 2006. Ethnomedicinal uses of plants by the Sarania tribe of Nalbari district of Assam. *J. Econ. Taxon. Bot.* 30(Suppl.): 133-139.

Sasidharan, N. & Augustine, Jomy. 2006. Ethnobotany of the tribes living in and around the Periyar Tiger Reserve, Southern Western Ghats, India. *J. Econ. Taxon. Bot.* 30(Suppl.): 45 - 58.

Satapathy, K. B. 1992. Medicinal uses of some Plants among the tribals of Sundargarh district, Orissa. *J. Econ. Taxon. Bot. Addl. Ser.* **10:** 241-249.

Sattapathy, K. B. & M. Brahmam. 2003. Some interesting phytotherapeutic claims of tribals of Jajpur district, Orissa. 467-472. In: Maheshwari, J. K. (ed.). Ethnobotany and Medicinal Plants of Indian Subcontinent. Scientific Publishers (India), Jodhpur.

Saxena, H.O. & P.K. Dutta. 1975. Studies on the ethnobotany of Orissa. *Bull. Bot. Surv. India.* 17: 124 – 131.

Saxena, A.P. & K.M. Vyas. 1981. Ethnobotanical records on infections diseases from tribals of Banda district. *J. Econ. Taxon. Bot.* 2: 191 – 194.

Saxena, H.O., M. Brahman & P.K. Dutta. 1931. Ethnobotanical studies in Orissa, 232 – 244. In S.K. Jain (Ed.). *Glimpses of Indian Ethnobotany.*

Saxena, A.P. & K.M. Vyas. 1983. Ethnobotany of Dhasan Valley. *J. Econ. Taxon. Bot.* 4: 121 – 128.

Saxena, A.P. & K.M. Vyas. 1986. Antimicrobial activity of seeds of some ethnomedicinal plants. *J. Econ. Taxon. Bot.* 8: 291 – 299.

Saxena, H.O. 1986. Observations on the ethnobotany of Madhya Pradesh. *Bull. Bot. Surv. India.* 28: 149 – 156.

Samvatsar, Swati. 1996. The flora of western tribal Madhya Pradesh, *Scientific Publisher,* Jodhpur.

Sebastian, M.K. & M.M. Bhandari. 1984. Medico-ethnobotany of Mount Abu, Rajasthan. India. *J. Ethnopharmacol.* 12: 223 – 230.

Sebastian, M.K. 1984. Plants used as veterinary medicines, galactagogues and fodder in the forest areas of Rajasthan. *J. Econ. Taxon. Bot.* 5: 785 – 788.

Sebastian, M.K. & Bhandari, M.M. 1984. Some plants used as veterinary medicines by bhils. *Int. J. Trop. Agric.* 2: 307-310.

Sen, R., D. C. Pal. & B. Roy. 1984. Botany and ethnobotany of Parashi. *J. Econ. Taxon. Bot.* 5: 857 – 858.

Sen, R., P. Sur, S. Bandyopadhyay & D.C. Pal. 1984. Ethnobotanical uses of Herbaria-III. *J. Econ. Taxon. Bot.* 5: 795-799.

Sen, R.S. Bandopadhayay, E. Biswas & A. Guha. 1985. Ethnobotanical uses of Herbaria – IV. *J. Econ. Taxon. Bot.* 6: 331 – 335.

Sen, S.K. & Pradhan, N.B. 1999. Conservation of ethnomedicinal plants of Bargarh district in Orissa. *Adv. Plant Sci.* 12: 207-213.

Shah, N.C. & M.C. Joshi. 1971. An ethnobotanical study of Kumaon region of India. *Econ. Bot.* 25: 414 – 422.

Shah, G.L., A.R. Menon & G.V. Gopal. 1981. An account of the ethnobotany of Saurashtra in Gujarat state (India). *J. Econ. Taxon. Bot.* 2: 173 – 182.

Shah, N.C. 1982. Herbal folk medicines in northern India. *J. Ethnopharmacol.* 6: 293-301.

Shah, G.L. & Gopal, G.V. 1982. An ethnobotanical profile of the Dangies. *J. Econ. Taxon. Bot.* 3: 355-364.

Shah, G.L., S.S. Yadav & N. Badri, 1983. Medicinal plants from Dahanu Forest Division in Maharashtra State. *J. Econ. Taxon. Bot.* 4: 141 – 151.

Shah, G.L. 1984. Some economically important plants of Salsette Island near Bombay. *J. Econ. Taxon. Bot.* 5: 753 – 765.

Shah, G.L. & G.V. Gopal. 1986. Folklore medicines of Vasavas, Gujarat, India. *Acta. Bot. Indica.*14: 48 – 53.

Shah, G.L. & Gopal, G.V. 1992. An ethnobotanical profile of the Dangies. *J. Econ. Taxon. Bot.* 3: 355 – 364.

Shah, M. A. U. 2005. Ethnomedicinal plants of Bhakkar Tehsil in Punjab, Pakistan. *Ethnobotany.* 17: 171-175.

Sharma, R. K. and Bhagvan Dash. 1976. Charaka samhita, vol. I & II, Chaukhamba Sanskrit Series, Varansi, India.

Sharma, P.K., S.K. Dhyani & V. Shanker. 1979. Some useful and medicinal plants of the district Dehradun and Siwalik. *J. Sci. Res. Pl. Med.* 1: 17 – 43.

Sharma, B.D. & S.K. Malhotra. 1984. A contribution to the ethnobotany of tribal areas in Maharashtra. *J. Econ. Taxon. Bot.*5: 533 – 537.

Sharma, K. K. & Shiva Sharma. 1992. Medicinal plants of Southern Aravalli Hills. *J. Econ. Taxon. Bot. Addl. Ser.* **10:** 209-217.

Sharma, P.V. 2003. Dravyaguna - Vijnana. Chaukhamba bharti academy Varanasi.

Sharma, B. D. & J. C. Rana 2003. Traditional medicinal uses of plants of Himachal Hills. 173-176. In: Maheshwari, J. K. (ed.). Ethnobotany and Medicinal Plants of Indian Subcontinent. Scientific Publishers (India), Jodhpur.

Sharma, H. M., A. Radhapyari Devi & B. Manihar Sharma. 2003. Ethnomedicinal uses of Monocotyledonous plants by the Meiteis of Manipur. 473-480. In: Singh *et al.* (ed.). Ethnobotany and Medicinal Plants of India and Nepal. Vol-I. Scientific Publishers (India), Jodhpur.

Sharma, P. K., N. S. Chauhan & Brij Lal. 2003. Comercially important medicinal and aromatic plants of Parvati valley, Himachal Pradesh. 937-942. In: Singh *et al.* (ed.). Ethnobotany and Medicinal Plants of India and Nepal. Vol-II. Scientific Publishers (India), Jodhpur.

Sharma, U.K. 2003. Folk and herbal medicine among Nepalese of Assam. 599-603. In: Maheshwari, J. K. (ed.). Ethnobotany and Medicinal Plants of Indian Subcontinent. Scientific Publishers (India), Jodhpur.

Sharma, S. C., G.N. Sharma & R.K. Aggarwal. 2006. Traditional significant Medicinal plants of Nagaur district (Rajasthan). *J. Econ. Taxon. Bot.* 30(Suppl.): 320-325.

Shekhawat, G.S. & Sunil Anand. 1984. An ethnobotanical profile of Indian Desert. *J. Econ Taxon. Bot.* 5: 591 – 598.

Shubhangi Pawar & D.A. Patil. 2006. Ethnomedicinal aspects of naturalised exotic plants of Jalgaon district, Maharashtra, India. *J. NTFPs* **13(1)**:55-59.

Shukla, G., V. Mudgal & K.K. Khanna. 1992. Notes on medicinal uses of plants known amongst rural folk-lore of Pratapgarh and Sultanpur districts, U.P. *J. Econ. Taxon. Bot. Addl. Ser.* **10:** 219-225.

Shukla, G., B.K. Verma & H.P. Pandey. 2003. Folk-lore herbotherapy of Jalaun district, U.P. India. 143-147. In: Singh *et al.* (ed.). Ethnobotany and Medicinal Plants of India and Nepal. Vol-I. Scientific Publishers (India), Jodhpur.

Siddiqui, M.B. & Husian, W. 1991. Traditional treatment of diarrhoea and dysentery through herbal drugs in rural India. *fitoterapia.* 62: 325 – 329.

Siddque, M. A.A., Jhon, A. Q. & Paul, T.M. 1995. Status of some important medicinal and aromatic plants of Kashmir Himalaya. *Adv Plant Sci.* 8: 134-139.

Sikarwar, R.L.S. & Kaushik, J.P. 1993. Folk medicines of the Morena district, M.P., India. *Int J Pharmacog.* 31: 283-287.

Singh, Harbhajan. 1945. Home useful wild plants of the Delhi Province. *India. J. Agr. Sci.*15: 297 – 307.

Singh, K.K., S.K. Palvi & H.B. Singh. 1980. Survey of some medicinal plants of Dharchula block in Pithoragarh district of U.P. *Bull. Medico-ethnobot. Res.*1: 1– 7.

Singh, V. & R.P. Pandey. 1980. Medicinal plantlore of the tribals of eastern Rajasthan. *J. Econ. Tuxon. Bot.*1: 137 – 147.

Singh, K.K. & J.K. Maheshwari. 1983. Traditional phytotherapy amongst the tribals of Varanasi district, Uttar Pradesh. *J. Econ. Tuxon Bot.*4: 829 – 838.

Singh, K.K. & Saha & J.K. Maheshwari. 1985. Ethnobotany of *Helicteres isora* L., in Kheri district, Uttar Pradesh. *J. Econ. Taxon. Bot.* 7: 487 – 492.

Singh, A.K., R.N. Singh & S.K. Singh. 1987. Some ethnobotanical plants of Terai region of Gorakhpur, district- I. *J. Econ. Taxon. Bot.* 9: 407 – 410.

Singh, H. 1988. Ethnobiological treatment of Piles by Bhoxas of Uttar Pradesh. *Anc. Sci. Life.* 8: 167 – 170.

Singh, L.B., A.K. Verma & S.S.N. Sinha. 1992. Preliminary observations on the Ethnomedicinal plants of Godda district (Bihar). *J. Econ. Taxon. Bot. Addl. Ser.* **10:** 205-208.

Singh, H. & Maheshwari, J.K. 1992. Traditional remedies for snake bite & Scorpion – sting among the Bhoxas of Nainital district. U.P. *Aryavaidyah.* 6: 120 123.

Singh, L.B. & Singh, C.L. 1992. An Ethno-Medico-botanical study of Deoghar district (Bihar). *Biojournal.* 4: 83 – 86.

Singh, P.B. & Aswal, B.S. 1992. Medicinal plants of Himachal Pradesh used in Pharmaceutical Industry. *Bull. Med. Ethno. Res.* 13: 172 – 208.

Singh, J. P., Suchita Kumar, Th. Puinyabati Devi & Sudhir Kumar. 1992. Medicinal Plants of Manipur – I. *J. Econ. Taxon. Bot. Addl. Ser.* **10:** 233-239.

Singh, P.B. 1993. Medicinal plants of Ayurvedic importance from Mandi district of Himachal Pradesh. *Bull. Med. Ethnobot. Res.*14: 126 – 136.

Singh, K.K. and Prakash, A. 1994. Indigenous phytotherapy among the Gond tribes of Uttar Pradesh, India. *Ethnobotany*.6: 37 – 41.

Singh, V. 1996. Ethnomedico-botany of Dards tribe of Gurez valley of Kashmir Himalaya. In S. K. Jain (Ed.): Ethnobiology in human welfare. Deep Publication, Delhi. pp.129-132.

Singh, K.K. 1997. Studies on native medicine of Jaunsari tribe of Dehradun district, U.P. India. *Int. J Pharmacog*.35: 105 – 110.

Singh, V.K., Ali, Z.A., & Siddique, M.K. 1997. Folk medicinal plaats of the Garwal and Kumaon forest of U.P. India. *Hamdard Med* 40(4): 35-47.

Singh, H.B.K. 1997. Studies on Medico-botany of Meitel community in Manipur State, India (II). *Adv. Plant. Sci*.10: 13 – 18.

Singh, V.K., Ali, Z.A. & Siddiqui, M.K. 1997. Medicinal plants used by the forest ethnics of Gorakhpur district (U.P.). *Int. J. Pharmacog*.35: 194 – 206.

Singh, S.K., Das. P.K. & Das, M.N. 1998. Ethnobotanical studies on some medicinal plants of Rajgir, Bihar. *Int. J. Mendeol*.15: 145 – 148.

Singh, V. & R. P. Pandey.1998. Ethnobotany of Rajasthan. Scientific publisher, Jodhpur.

Singh, G. S. 2003. Ethnobotanical study of useful plants of Kullu district in North Western Himalaya. 185-198. In: Maheshwari, J. K. (ed.). Ethnobotany and Medicinal Plants of Indian Subcontinent. Scientific Publishers (India), Jodhpur.

Singh, A. K. 2003. A contribution to the ethnobotany of Sub-Himalayan region of Eastern U. P. 237-246. In: Maheshwari, J. K. (ed.). Ethnobotany and Medicinal Plants of Indian Subcontinent. Scientific Publishers (India), Jodhpur.

Singh, G. S. 2003. Ethnobotanical inventory of Sariska national park in the Aravalli Hills of Eastern Rajasthan. 181-195. In: Singh *et al.* (ed.). Ethnobotany and Medicinal Plants of India and Nepal. Vol-I. Scientific Publishers (India), Jodhpur.

Singh, H. 2003. Herbal recipes for spermatorrhoea by Bhoxa tribe of Uttaranchal. *Ethnobotany*. 15:115-117.

Singh, P. K. & K. Ibotombi Singh. 2003. Mother and Child birth: I- An ethnobotanical study of the Meitei community of Manipur State. 457-465. In: Singh *et al.* (ed.). Ethnobotany and Medicinal Plants of India and Nepal. Vol-I. Scientific Publishers (India), Jodhpur.

Singh, P. K. & K. Ibotombi Singh. 2003. First-aid remedies: An ethno-medico-botanical study of the Meitei community of Manipur State. 466-472. In: Singh *et al.* (ed.). Ethnobotany and Medicinal Plants of India and Nepal. Vol-I. Scientific Publishers (India), Jodhpur.

Singh, K. K. & K. Kumar. 2003. Ethnotherapeutics of some Medicinal plants used as antipyretic agents among the tribals of India. 135-141. In: Maheshwari, J.K. (ed.). Ethnobotany and Medicinal Plants of Indian Subcontinent. Scientific Publishers (India), Jodhpur.

Singh, V. P. & R. C. Srivastava. 2003. Economic importance of Sedges of Uttar Pradesh. 978-981. In: Singh *et al.* (ed.). Ethnobotany and Medicinal Plants of India and Nepal. Vol-II. Scientific Publishers (India), Jodhpur.

Singh, S.P., S. Tripathi & R.S. Shukla, 2003. Ethnomedicinal heritage for Bioprospecting and drug development in North-Eastern States of India. 384-395. In: Singh *et al.* (ed.). Ethnobotany and Medicinal Plants of India and Nepal. Vol-I. Scientific Publishers (India), Jodhpur.

Sinha, B. K., B. K. Shukla & P. Sharma. 2006. Ethnomedicinal plants of Chhatisgarh with sporadic distribution. *J. Econ. Taxon. Bot.* 30(Suppl.):70-109.

Sivakumar, A., M. S. Subramanian, M. Karunakaran & A. Burkanudeen 2003. Ethnobotany of Poliyars of Anaimalai Hills, Tamil Nadu. 679-685. In: Singh *et al.* (ed.). Ethnobotany and Medicinal Plants of India and Nepal. Vol-II. Scientific Publishers (India), Jodhpur.

Sood, S. K. & S. thakur. 2004. Ethnobotany of Rewalsar Himalaya. Deep Publications. New Delhi.

Srivastava, D.K. & S.K. Varma. 1981. An ethnobotanical study of Santhal Pargana, Bihar, *Inidan. Forester.* 107: 30 – 41.

Srivastava, T. N., Rajasekhara, S., Badola, D. P. & Shah, D. C. 1986. An index of the available medicinal plants, used in Indian system of medicine from Jammu & Kashmir State. *Anci. Sci. Life.* 6: 49-68.

Srivastava, P. K., K. K. Khanna & V. Mudgal. 1992. New Traditional herbal remedies from the rural folklore of Hamirpur district, U. P. *J. Econ. Taxon. Bot. Addl. Ser.* **10:** 399-404.

Srivastava, S.C. and Rout, N. 1994. Some plants of ethnopaediatric importance in district Koraput, Orissa. *Bull. Bot. Surv. India.* 36: 166 – 168.

Srivastava, R. C., V. P. Singh & M. K. Singh. 2003. Medicinal plants of Jaunpur district U. P. India. 148-159. In: Singh *et al.* (ed.). Ethnobotany and Medicinal Plants of India and Nepal. Vol-I. Scientific Publishers (India), Jodhpur.

Srivastava, A. K. & Monika Singh. 2003. Medicinal potential of grasses in Rajasthan. 170-176. In: Singh *et al.* (ed.). Ethnobotany and Medicinal Plants of India and Nepal. Vol-I. Scientific Publishers (India), Jodhpur.

Srivastava, S. C., S. Mitra & S. Bandyopadhyay. 2003. Some plants of ethnomedicinal importance from Darjeeling district, West Bengal. 972-977. In: Singh *et al.* (ed.). Ethnobotany and Medicinal Plants of India and Nepal. Vol-II. Scientific Publishers (India), Jodhpur.

Subramani, S. P. & G. S. Goraya. 2003. Some folklore medicinal plants of kolli Hills record of a Natti Vaidyas Sammelan. 665-678. In: Singh *et al.* (ed.). Ethnobotany and Medicinal Plants of India and Nepal. Vol-II. Scientific Publishers (India), Jodhpur.

Subramaniam, A. 2003. A survey of medicinal plants from Chitheri hills in Dharamapuri district, Tamilnadu. 395-416. In: Maheshwari, J. K. (ed.). Ethnobotany and Medicinal Plants of Indian Subcontinent. Scientific Publishers (India), Jodhpur.

Subramanian, A., V. R. Mohan, S. Kumaresan & A. Chelladurai. 2003. Medicinal plants used by the valaiyans of Madurai district, Tamil Nadu. 785-787. In: Singh *et al.* (ed.). Ethnobotany and Medicinal Plants of India and Nepal. Vol-II. Scientific Publishers (India), Jodhpur.

Sudhakar, S. & R.S. Rolla. 1985. Medicinal plants of Upper East Godavari District, Andhra Pradesh, and need for establishment of Medicinal Farm. *J. Econ. Taxon. Bot.* 7: 399 – 406.

Sudhakar, A. & Chetty, K.M. 1998. Medicinal importance of some angiospermic weeds used by the rural people of Chittoor district of Andhra Pradesh, India. *Fitoterapia.* 69: 390 400.

Sur, P. R., R. Sen & A. C. Halder. 1992. Ethnobotanical study of Purulia district, West Bengal. *J. Econ. Taxon. Bot. Addl. Ser.* **10:** 259-264.

Tarafder, C. R. 1978. Some interesting new uses of the well known plant *Vitex negundo* L. (NISINDA). Bull. Bot. Surv. India 20: 176-177.

Tarafder, C.R. & H.N., Rai Chaudhuri. 1981. Less known medicinal uses of plants among the tribals of Hazaribagh district of Bihar. 208 – 217. In S.K. Jain (Ed.). *Glimpses of Indian Ethnobotany.*

Tarafder, C.R. 1985. *Acacia catechu* (L.f.) Willd. (Fabaceae) - Inddigenous and less known uses of the plant. *Folklore.* 26: 158 – 159.

Tarafder, C.R. 1983. Traditional medicinal plants used by the tribals of Ranchi and Hazaribagh district, Bihar- Plants used in stomach troubles. *J. Econ. Taxon. Bot.* 4: 891 – 896.

Tarafder, C.R. 1983a. Ethnogynaecology in relation to plants – II. Plants used in abortion. *J. Econ. Taxon. Bot.* 4: 507 – 516.

Tarafder, C.R. 1983b. Ethnogynaecology in relation to plants – I. Plants used in antifertility and conception. *J. Econ. Taxon. Bot.* 4: 483 – 490.

Tarafder, C.R. 1986. Ethnobotany of Chhotanagpur (Bihar). *Folklore.* 27: 95 – 100.

Tarafder, C.R. 1987. Some traditional knowledge about tribal health. *Folklore.*28: 37–42.

Tarafder, C.R. 1984a. Ethnogynaecology in relation to plants. Part- III. Plants used to accelerate delivery & in pre & post-natal complaints. *J. Econ. Taxon. Bot.* 5: 572-576.

Tarafder, C.R. 1984b. Less known nine medicinal plants used by the tribals for curing gonorrhoea in Ranchi and Hazaribagh district, Bihar. *Folklore.* 25: 47- 49.

Tarafder, C.R. 1984c. Less known twenty three medicinal plants used by the tribals for curing boils in Ranchi and Hazaribagh district. Bihar. *Vanyajati.* 32: 14 – 19.

Tewari, L.C., Agrawal, R.G. Pandey, M.J. Uniyal, M.R. & Pandey, G. 1990. Some traditional folk medicine from the Himalayas (U.P. region), *Aryavaidyan* 4: 49-57.

Tiwari, K. C., R. Majumdar & S. Bhattacharjee. 1980. Tribal medicines and treatment from distict siang of Arunachal Pradesh. *Bull. Medico-ethnobot. Res.* 1: 440-444.

Thomas, Jery & Jhon De Britto. 2003. Weeds of medicinal importance in Tirunelveli district in Tamil Nadu. 363-367. In: Maheshwari, J.K. (ed.). Ethnobotany and Medicinal Plants of Indian Subcontinent. Scientific Publishers (India), Jodhpur.

Thomas, J. & A. John De Britto. 2003. Ethnobotanical study in Wynad district in Kerala. 815-824. In: Singh *et al.* (ed.). Ethnobotany and Medicinal Plants of India and Nepal. Vol-II. Scientific Publishers (India), Jodhpur.

Tigga, M., G. S. Lakra & P. S. N. Rao. 2003. Medicinal value of latex yielding plants from Andman & Nicobar Islands. 765-772. 60. In: Singh *et al.* (ed.). Ethnobotany and Medicinal Plants of India and Nepal. Vol-II. Scientific Publishers (India), Jodhpur.

Tiwari, K.C., R. Majumdar & S. Bhattacharjee. 1979. Folklore medicine from Assam and Arunachal Pradesh (district Tirap). *Quart. J. Crude Drug Res.* 17: 61 – 67.

Tiwari, K.C. & Majumdar, R. 1996. Medicinal plant form upper Assam borders having specific folk uses. *Sachitra Ayurved* 49: 207-214.

Tomar, A. & H. Singh 2006. Exotic medicinal plants from Baghpat district, Uttar Pradesh. India. *J. NTFPs* **13(4)**:273-280.

Tribedi, G.N., R.N. Kayal & H.N. Rai Chaudhuri. 1982. Some medicinal plants of Mayurbhanj (Orissa). *Bull. Bot. Surv. India.* 24: 117 – 120.

Tribedi, G.N., H.A. Molla & D.C. Pal. 1985. Some uses of plants from the tribal areas of Chotanagpur, Bihar. *Nagarjun. 29*: 15 – 18.

Tripathi, Y.C., Prabhu, V.V., Pal, R.S. & R.N. Mishra 1996. Medicinal plants of Rajasthan in Indian system of medicine. *Ancient Sci. Life.* 15: 190 – 212.

Udayan, P. S., S. George, K. V. Tushar, S. Sudhakar Raja & I. Balachandran 2006a. A note on the Ethnomedicinal plants used by the Kaadar Tribes, Kariyan Shola, Top Slip, Tamil Nadu. *J. Econ. Taxon. Bot.* 30(Suppl.): 5 – 12.

Udayan, P. S., S. George, K. V. Tushar, S. Sudhakar Raja & I. Balachandran. 2006 b. Ethnobotanical information from the Muduga tribe of Mukkali, near Silent Valley National Park, Palakkad district, Kerala. *J. Econ. Taxon. Bot.* 30 (Suppl.): 27 –30.

Uniyal, M.R. 1968. Medicinal plants of Bhagirathi valley lying in Uttarkashi forest division. *Indian Forester.* 94: 407 – 420.

Uniyal, M.R. & M.S. Chauhan. 1973. Traditionally important medicinal plants of Kangra Valley in Dharamsala forest circle, Himachal Pradesh. *J. Res. Indian. Med.* 8: 76 – 85.

Upadhye, A., Kumbhojkar, M.S. & Vartak, V.D. 1986. Observations on wild plants used in folk medicine in the rural areas of the Kolhapur district. *Ancient Sci. Life.* 6: 119-121.

Upadhyay, R. & Jaswant Singh. 2005. Ethnomedicinal uses of plants from Tikri forest of Gonda district. *Ethnobotany.* 17: 167– 170.

Upadhye, A. M.S. Kumbhojkar & V.D. Vartak. 1986. Observations on wild plants used in folk medicine in the rural areas of the Kolhapur district. *Anc. Sci. Life.* 6: 119 – 121.

Upadhay, O.P., Kumar, K. & Tiwari, R.K. 1998. Ethnobotanical study of skin treatment uses of medicinal plants of Bihar, *Pharmaceut Biol.* 36: 161 – 172.

Upadhyay, R. & S. V.S. Chauhan. 2003. Ethnobotanical uses of plant gums by the tribals. 601-602. In: Singh *et al.* (ed.). Ethnobotany and Medicinal Plants of India and Nepal. Vol-II. Scientific Publishers (India), Jodhpur.

Vargese SVD. 1996. Applied ethnobotany-A case study among the Kharias of central India. *Deep publications.* New Delhi.

Venugopal, K. & K. Raghunathan. 1982(Ed.). Medicobotanical-cum-ethnomedicinal survey of district Jhabua (M.P.). *Centr. Counc. Res. Ayurv. Siddha Newsletter.* 4 - 5.

Ved Prakash & S.K. Jain. 1984. Poaceae: Tribe - Isachneae. *Fasc. Fl. India.*14: 41 – 42. Bot. Surv. India. Calcutta.

Ved Prakash & B.N. Mehrotra. 1987. Ethnobotanical studies on the flora of Khandala, Maharashtra State. *J. Econ. Taxon. Bot.* 9: 205 – 208.

Ved Prakash & B.N. Mehrotra. 1988. Unrecorded traditional medicines-I. *Anc. Sci. Life.* 3: 110-112.

Verma, D.M. Balakrishan, N.P. & Dixit, R.P. 1993. Flora of Madhya Pradesh. BSI, Calcatta.

Verma, S. & N. S, Chauhan. 2006. Studies on ethno-medico-botany of Kunihar forest division, District Solan (H. P.). *Ethnobotany*18: 160-165.

Vijaya Kumar, R. & Pullaiah, T. 1998. Medicinal plants used by the tribals of Prakasham District, A.P. *Ethnobotany.* 10: 97-102.

Vir Jee, G.H. Dar, P. Kachroo & G.M. Bhat. 1984. Taxo – ethnobotanical studies of the rural areas in district Rajouri (Jammu). *J. Econ. Taxon. Bot.* **5:** 831 – 838.

Viswanathan, K. 1995. Survey on medicinal spices of the Nilgiris. *Ancient Sci. life.* 14: 258-267.

Warrier, P.K., V.P.K. Nambiar & C. Ramankutty. 1994. Indian medicinal plants. Vol. I-V. Oriental Longum, Chennai.

Yesodharan, K. & K.A. Sujana. 2006. Ethnomedicinal plants used by the trbals of Parambikulam Wildlife Sanctuary, Kerala to cure cuts and wounds. *J. Econ. Taxon. Bot.* 30(Suppl.): 365-369.

INDEX OF ENGLISH, HINDI & SANSKRIT NAMES OF PLANTS

Botanical Name	English Name	Hindi Name	Sanskrit Name	Page No.
Abroma angusta	Devil's cotton	Ulatkambal	Pivari	1
Acacia farnesiana	Cassie flower	Vilayati kikar	Irimeda	2
Acacia pennata	-	Aila	Ari	3
Acacia polyantha	White cutch	Svetakhair	Kadarah	4
Achillea millefolium	Yarrow	Gandana	Svetadurva	5
Albizia procera	White Sirish	Safed Siris	Kinihi	6
Alhagi camelorum	Camel thorn	Bharbharra	Yavasa	7
Alocasia indica	Giant taro	Mankanda	Manaka	8
Althaea officinalis	Marsh mallow	Gandhali	Prasarani	9
Altingia excelsa	-	Silaras	Silhaka	10
Amorphophallus sylvaticus	-	-	Lohita-surana	10
Ampelocissus latifolia	Jungle grape vine	Panibel	-	11
Angelica glauca	Long wort	Churai	Coraka	12
Anisomeles indica	-	-	-	13
Artemisia maritima	Worm seed	Ajavayana	Yavani	14
Artocarpus communis	Bread fruit	Bilatti Kathal	Ksudra panasah	15
Artocarpus hirsutus	Wild jack	Katahal	Lakucah	15
Arundo donax	Spanish cane	Bara nal	Nala	16
Atalantia monophylla	Wild lime	Jungli nimbu	Atavijambira	17
Brassica nigra	Black mustard	Rai	Asuri	18
Caesalpinia digyna	Teri pods	Vakerimul	Vakeri	19
Cajanus cajan	Pigeon pea	Tuvar	Adhaki	19
Calamus rotang	Common rattan	Bemt	Vetrah	20
Canavalia gladiata	Borad bean	Lalkudusumpal	Asisimbi	21
Canscora decussata	Canscora	Samkhaphuli	Sankhini	22
Canthium parviflorum	Carray cheddie	Kirma	Gangeruki	23

Capparis zeylanica	Ceylon caper	Ardanda	Karambha	24
Carissa carandas	Christ's thorn	Karaunda	Avighnah	25
Cassia alata	Ringworm Senna	Dadmurdan	Dadrughna	26
Cassia senna	Indian senna	Sanay	Svarnapatri	27
Cerbera odollam	Odallum tree	Dabur	Auddalakah	28
Chenopodium anthelminticum	Worm seed	Khatu	Svetachilli	29
Chonemorpha fragrans	-	Garbhedaro	Murva	30
Chrozophora rottleri	Turnsole	-	Suryavarta	31
Cicca distichia	Country gooseberry	Chalmeri	Lavali	32
Citrus aurantifolia	Sour orange	Khatta	Brhat jambirah	33
Clematis triloba	-	Churahar	Morata	34
Colchicum luteum	Colchicum	Hirantutiya	Hiranyaththa	34
Coleus vettiveroides	-	Valak	Valakam	35
Commelina benghalensis	Day flower	Kankauoa	Kanchata	36
Commiphora myrrha	Myrrh	Bol	Bolah	36
Convolvulus arvensis	Dear's foot	Harin padi	Rajbali	37
Corchorus olitorius	Jew's mallow	Koshta	Pattshak	38
Coscinium fenestratum	Tree turmeric	Jhade-halde	Darvi	39
Crinum defixum	-	-	Vishamandala	39
Croton oblongifolius	-	Chucka	Bhutankusa	40
Cucumis melo var. *utilissimus*	Snake cucumber	Kakkri	Karkati	41
Cucumis sativus	Cucumber	Khira	Trapusah	42
Cucumis trigonus	Bitter gourd	Jangli Indrayan	Indravaruni	43
Cucurbita maxima	Red gourd pumpkin	Kaddu	Pitakusmandah	44
Cyamopsis tetragonoloba	Cluster bean	Govar	Ksudrasimbi	45
Cyathula prostrata	Small prickly chaff-flower plant	Lal cicimda	Raktapamargah	46
Cyclea peltata	Pata root	Path	Patha	47
Cyperus scariosus	-	Nagarmutha	Bhadramusta	48
Delphinium denudatum	-	Nirbisi	Nirvisa	49
Dendrocalamus stirctus	Solid bamboo	Bans	Vansha	50
Desmodium triflorum	-	Kudaliya	Tripadi	50
Digera arvensis	-	Latmahuria	Aranya vastuka	51
Digitalis purpurea	Foxglove	Digitalis	Tilapuspi	52
Diospyros melanoxylon	Coromandel ebony persimmon	Tendu	Dirghpatraka	52

Diospyros malabarica	Gaub Persimmon	Kalatendu	Tinduka	53
Emilia sonchifolia	-	Hirankhuri	Sasasrutih	54
Euphorbia thymifolia	Didiya	Dudhi	Dugdhika	55
Ferula narthex	Tibetian asafoetida	Hing	Sahastravedhi	56
Ficus arnottiana	-	Paras pipal	Plaksha	57
Flacourtia jangomas	Many spiked flacorita	Talispatri	Prachinamalaka	57
Garuga pinnata	Garuga	Khogar	Krsnamlika	58
Gentiana kurroo	Indian gentian	Trayaman	Trayamana	59
Grewia tilifolia	Dhaman	Dhaman	Dhanvangah	60
Habenaria edgeworthii	-	Rddhi	Rddhi	61
Habenaria intermedia	-	Vrddhi	Vrddhi	62
Hedyotis herbacea	-	Paper-bhed	Chayaparpatika	62
Hedychium spicatum	Spiked ginger lily	Kapurkachri	Kapurakachali	63
Heliotropium indicum	Indian turnsole	Siriyari	Vrscikali	64
Helleborus niger	Black hellebore	Kalikatuki	Katuki	65
Hibiscus aculeatus	-	-	Sathambasthi	66
Hibiscus cannabinus	Mesta	Pitwa	Anvastha	66
Holostemma ada-kodien	-	Chirvel	Jivanti	67
Homonoia riparia	-	Jalbent	Jalavetasah	68
Hugonia mystax	Climbing flax	-	Kamsamarah	69
Hygroryza aristata	Bengal wild rice	Jangali-Dal	Nivarah	70
Illicium verum	Star anise of China	Anasphal	Takkolakam	70
Ipomoea pes-caprae	Goat's foot creeper	Dopatilata	Maryadavalli	71
Ipomoea sepiaria	-	Bankalmi	Laksmana	72
Ixora coccinea	Jungle-flame ixora	Rangan	Paranti	73
Jasminum angustifolium	Wild Jasmine	Banmallika	Asphota	74
Jasminum arborescens	Tree jasmine	Bela	Vasanti	74
Jasminum auriculatum	Needle flower Jasmine	Juhi	Yuthika	75
Jasminum multiflorum	Downy jasmine	Kundphul	Kundah	76
Justicia beddomei	-	Adusa	Vasa	77
Kaempferia galanga	-	Kandramula	Sathi	77
Lagerstroemia speciosa	Pride of India	Jarul	Tinisa	78

Launea coromandelica	Wodier	Jhingan	Jhingini	79
Leea indica	-	Kurkurjihva	Chatri	80
Leptadenia reticulata	-	Dori	Jivanti	81
Lilium polyphyllum	Ksheerakakoli	Ksirakakoli	Ksirakakoli	81
Limnophilla aromatica	-	Kutra	Amragandhah	82
Liquidamber orientalis	Oriental sweet-gum	Silaras	Sihlakah	83
Macrotyloma uniflorum	Horse gram	Kulthi	Kutatthah	84
Malaxis muscifera	Risabak	Rsabhak	Rsabhakah	85
Manihot esculenta	Tapioca	Sakkarkand	Kalapakandah	85
Maranta arundinacea	Arrowroot	Tikhor	Tavaksiri	86
Martynia annua	Devil's claw	Bichua	Kaknasa	87
Mentha spicata	Garden mint	Podina	Putiha	88
Merremia tridentata	-	-	Prasarani	89
Monochoria vaginalis	-	-	Indivarah	89
Morinda pubescens	Morinda tree	Acc	Paphanah	90
Mussaenda frondosa	White calycine	Bedina	Rajatarih	91
Myxopyrum serratulum	-	Pilacameli	Hemamalati	92
Naravelia zeylanica	-	-	Dhanavalli	93
Nervilia aragoana	-	Sthalapadma	Padmacarini	93
Nilgirianthus ciliatus	-	Karvi	Sahacarah	94
Ocimum kilimandscharicum	Camphor basil	Kapurtulsi	Karpuratulasi	95
Ophiorrhiza mungos	Mongoose plant	Sarhati	Sarpari	96
Opuntia elatior	Prickly pear	Naagaphanaa	Kantaseva	96
Orchis latifolia	Marsh orchid	Salam	Salampamisri	97
Origanum majorana	Sweet majoram	Marua	Marubaka	98
Oryza sativa	Paddy	Chawal	Syali	99
Ougeinia oojeinensis	Sandan	Sandan	Tinisah	100
Oxystelma esculentum	-	Dudlata	Dughika	101
Panicum sumatrense	Little millet	Savan	Syamakah	102
Paspalum scrobiculatum	Punctured paspalum	Kodo	Hudala	102
Pentapetes phoenicea	Noon plant	Dopahariya	Bandhukah	103
Phaseolus mungo	Black gram	Urid	Masha	104
Phoenix pusilla	Small wild date-palm	Palavat	Parusakah	105
Phyllanthus reticulatus	-	Pancoli	Krsnakamboji	105

Piper brachystachyum	-	Cab	Cavyah	106
Plumbago indica	-	Chitra	Chitraka	107
Podophyllum emodi	Indian podophyllum	Bakrachimaka	Hansapadi	108
Polygonatum cirrhifolium	Meda	Meda	Meda	109
Polygonatum verticillatum	Mahameda	Mahameda	Mahameda	110
Prunus armeniaca	Common apricot	Chuari	Aluka	111
Pseudarthria viscida	-	Salparni	Salparni	111
Pterospermum canescens	-	Mucukund	Mucukundah	112
Ranunculus sceleratus	Blister buttercup	Jal dhania	Kandira	113
Rhaphidophora laciniata	-	-	Sphotyabhujangam	114
Rhinacanthus nasuta	-	Palakjuhi	Gajakarni	114
Rhus parviflora	-	Tung	Tintidika	115
Rotula aquatica	-	-	Pasanabhedah	116
Rungia repens	-	Kharmor	Parpatha	117
Saccharum arundinaceum	Devil sugarcane	Ramsar	Munjah	117
Saccharum munja	-	Munj	Sara	118
Saccharum spontaneum	Thatch grass	Kus	Kasah	119
Salacia reticulata	-	-	Vairi	120
Salvadora persica L. var. *Wightiana*	*Tooth brush tree*	Jhak	Angahava	120
Samadera indica	Niepa bark tree	Lokhanti	Gucchakaranjah	121
Sapindus laurifolia	Soapnut tree	Ritta	Phenilah	122
Sarcostemma acidum	Moon plant	Somlata	Somavalli	123
Sarcostigma kleinii	-	-	Ingudi	124
Sassurea lappa	Kuth	Kuth	Kusthah	125
Schleichera oleosa	Lac tree	Kusum	Mukulakah	126
Schrebera swietenioides	Weaver's beam tree	Moka	Muskakah	127
Scindapsus officinalis	-	Gajapipal	Gajapippali	128
Securinega leucopyrus	-	Aints	Bhuriphali	129
Semecarpus anacardium	Marking nut tree	Bhilawa	Bhallataka	129
Sesamum indicum	Gingelly	Til	Tilah	131
Setaria italica	Deccan grass	Bartiya	Kanghuni	132
Sida spinosa	Prickly Sida	Gangern	Bala	133
Sisymbrium irio	London-Rocke	Khubkalan	Khubakala	134
Solanum dulcamara	Bitter sweet	-	Kakmachi	134

Solena amplexicaulis	-	Anatmul	Amlavetasah	135
Sonchus arvensis	Hogweed	Sadhi Sadhi	-	136
Sorghum vulgare	Indian millet	Juar	Yavanala	137
Spondias pinnata	Wild mango	Junli-Am	Amratakah	138
Sterculia foetida	-	Jungalbadam	-	139
Sterospermum chelonoides	Trumpet flower	Padhal	Alivallabha	140
Stereospermum colais	Trumpet flower tree	Padaria	Aliballabha	141
Streblus asper	Siamese rough bush	Daheya	Akshadhara	142
Strychnos potatorum	Celaring Nut	Nirmali	Katakah	143
Syzygium jambos	Rose-apple	Gulab-Jamun	Campeyah	144
Tabernaemontana heyneana	East Indian rose bay	-	Sveta-kutaja	145
Taraxacum officinale	Common dandelion	Dudhal	Dudhapheni	145
Terminalia paniculata	Flowering murdah	-	Asvakarnah	146
Thalictrum foliolosum	-	Pilazari	Mamira	147
Tinospora malabarica	-	Giloe	Kandodbhava-guruci	148
Toddalia asiatica	Wild orange tree	Dahan	Dahana	149
Trachyspermum ammi	Carum	Ajwain	Ajawodika	150
T. roxburghianum	Ajowan	Ajmud	Ajoda	151
Tragia involucrata	Indian stinging Nettle	Barahanta	Dhumpushpa	152
Trianthema portulacastrum	Horse purslane	Sabuni	Chiratika	153
Trichodesma indicum	-	Chhota kulpa	Surasa	154
Trichosanthes cucumerina	-	Jungli chachinda	Amritahala	155
Trichosanthes lobata	Bitter snake gourd	Jangli cicinda	Patola	156
Trichosanthes tricuspidata	-	Lal indrayan	Mahakasa	156
Tridax procumbens	Coat buttons	-	-	157
Triticum aestivum	Bread wheat	Gehun	Arupa	158
Triumfetta rhomboidea	Spiny cockle burr	Chikti	Jhinjharita	159
Typha elephantina	Elephant grass	Mothitrina	Gundra	160
Uvaria narum	-	-	Nilavalli	161
Valeriana jatamansi	Indian valerian	Tagar	Tagarah	162
Vateria idica	Piney varnish	Chandras	Ajakaran	163

Ventilago maderaspatana	Red creeper	Pitti	Raktavalli	164
Vigna pilosa	-	Banmung	Mudgaparni	165
Vigna radiata	Green gram	Mung	Mudgah	166
Vigna radiata (L.) Wilczek var. *Sublobata*	Wild black gram	Banurd	Masaparni	166
Vigna unguiculata	Cow pea	Sonta	Chavala	167
Wedelia chinensis	-	Bhangra	Bhringaraja	168
Zanonia indica	Bandolier fruits	Chirpoti	Kuntali	169
Ziziphus nummularia	Wild jujube	Jharberi	Kshudrabadar	169
Ziziphus oenoplia	Jackal jujube	Makai	Srigalakoli	170